composting with worms

why waste your waste?

by George Pilkington

eco-logic books

First Published in 2005, reprinted 2007 by
eco-logic books
Mulberry House, 19 Maple Grove, Bath, BA2 3AF

Web: www.eco-logicbooks.com

ISBN 1 899233 13 X

Illustrations by Susanna Kendall
Editing by Marjorie Gibbon
Design and Typesetting by Steve Palmer
Printed and bound in the UK by Russell Press
Printed on deinked post consumer waste - this book is fully compostable.

CONTENTS

Acknowledgements

There are many people who have either contributed directly to this book, or have inspired me over the years, and others who have offered advice and expertise. I would like to thank all of the following: Dr. Mark Langan of Manchester Metropolitan University, Nigel Baker of Worms Direct, Elaine Ingram of Soil Food Web and Dr. Trevor Piearce of Lancaster University.

Invaluable advice and assistance came from Kelly Slocum from the USA, And from the numerous expert contributors to the American web forum 'Worm Digest' especially Frank Teuton and Cheryl Paige, and also from Sylvia Hopwood, host of the UK worm forum. I gained great inspiration from Mary Appelhof – she will be sadly missed. Big thanks to Phil Hackett, Gregg Paget, Val Rawlinson and Hayley Holland for their support and for believing in my quest. And I'm grateful to Marjorie Gibbons who had the mammoth task of reducing the original and definitive 'magnum opus' to this more concise work!

Thanks also to Manchester Metropolitan University for the use of their research library and other facilities, without which the investigations for this book would have taken so much longer. Jim Hay for taking me to Wales, Peter Andrews (eco-logic books) for asking me to write the book and my family, who once again suffered with me in front of the computer, writing and researching for hours and hours!

And of course, a special thanks to the worms.

Introduction

There are many animals called 'worms', not only earthworms. However, throughout this book I will use the term 'worm' synonymously with earthworm

I was first inspired by the potential of worms in about 1986 when I read a book called Gardening Without Chemicals by Jack Temple. He mentioned in just a few lines how he had turned domestic waste into plant food using worms housed in plastic dustbins. As a result, I made my first worm bin.

Some time later, an organic author and gardening friend of mine, Jim Hay, contacted me to say that he was writing an article about using worms to compost kitchen waste. He had been invited to Turning Worms, a worm composting company in Wales, and asked if I would like to accompany him. I didso, and it inspired me. I made another wormery, this time from wood.

I wrote several articles about worm composting and gave numerous talks to gardening and horticultural societies throughout the North West of England and Wales. Later, when I started teaching about organic and wildlife gardening for adult education establishments, I incorporated these talks into my teaching. I was amazed that even when giving talks to organic gardening groups, the vast majority of attendees did not operate a wormery to recycle their food waste. Garden waste was composted, but food waste was, in nearly every case, literally binned and sent to landfill. Some years ago, I weighed my family's food waste generated between April to October. It came to 140 Kg. By composting that waste I saved it from going to landfill.

I was acutely aware that here in the UK there was very little information and experience in making wormeries. There was virtually nothing on how to set them up and manage them, in fact precious little public awareness or education on the topic. In the nineties, plastic wormeries were appearing but they did require certain management techniques to be successful. In too many cases people purchased a wormery, found there was just too much bother involved in managing them efficiently, and quietly gave up. Hardly a recipe for a successful vermicomposting movement.
Looking over the water at the USA, vermicomposting is a big business commercially and well established domestically. The worms found a champion in Mary Appelhof. Her highly successful book, Worms Eat My Garbage, was for many years the 'bible' of vermicomposting.

What we needed here in Europe and the UK was education, inspiration and well-designed wormeries that worked. For these reasons, a colleague and I set about designing the Waste Buster series of wormeries, which I then had manufactured,

I then decided to write this book - I hope it will inspire you to set up your own wormery and that you find keeping worms as satisfying as I do.

1 Why worm compost?

"I became deeply interested in the preparation of humus on a large scale; the place of the earthworm in converting crude vegetable matter into food materials for the crop appeared more and more important as the years passed"

Sir Albert Howard

The natural process of decomposition

The decay of organic matter into humus has taken place naturally for millennia. Wherever vegetation has grown and died, wherever an animal has died, decomposition takes place. This natural process for most purposes is slow and stable, and supports life on land, as we know it today. A circle of life, death, decay and life, all made possible by microbes - and, of course, the humble earthworm.

Decay in the wild

The largely mineral-based soil is perhaps the most complex of the many habitats in which microbes are found. In the soil of forests, fields, and gardens the microbes live alongside trees, flowers, worms, woodlice and a myriad of tiny organisms. When vegetation dies it falls on top of the soil where it accumulates and provides food for a variety of organisms and microbes. They break it down and eventually it forms humus, which is extremely important to the fertility and structure of soil.)

Hot Spots - the place to be!

There are regions within the soil that can be called 'hot spots' of microbial activity - the nightclubs of the soil! The area around plant roots, the rhizosphere, is an important hot spot. Microbes are very active around growing roots, where they feed on dead tissue, root hairs and root cells. Worms are attracted to this activity, finding lots to eat, and themselves increasing the waste and nutrient cycling. All this activity around the roots improves the soil structure, with waste materials and worm casts forming larger soil particles, improving aeration and drainage, resulting in greater root growth and increasing activity further. This can be seen in a worm bin with a window – I have grown potatoes in one and watched the worms congregate around the plant roots.

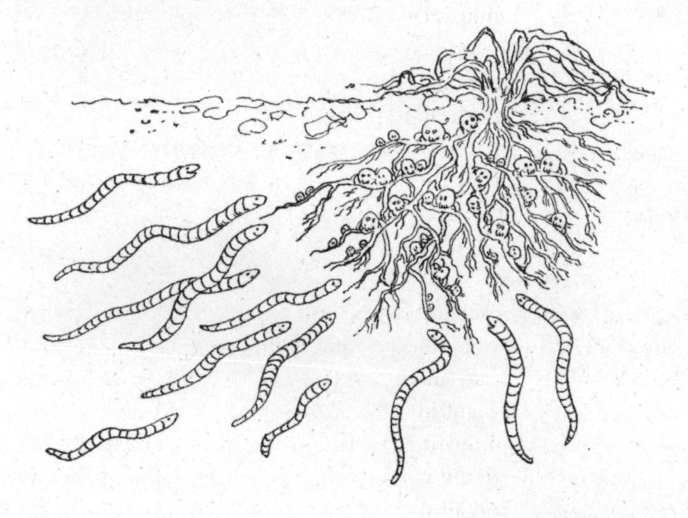

Other hot spots are around worm burrows and the area directly underneath decaying logs. Over 90% of the microbes and other soil life live in hot spots, although they make up less than 10% of the total soil volume.

Compost heaps

We tend to think of compost heaps as normal, nothing special, but we should remember that they are an an unnatural phenomenon. A man-made heap could be 2 metres (6 feet) high or more and contain grass cuttings, leaves, horse manure, twigs, newspaper, food waste, tea bags etc. Certain compounds that decompose in a compost heap may occur at rates that would never occur naturally.

Natural decay is a very complex system, but man has learnt to manipulate it to his advantage. While researching this book I came to realise how important manure and compost actually were to our ancestors. Without these natural fertilisers and soil conditioners we would be living in a very different world.

The decomposition process is much more rapid in a compost heap than in the wild. There are no growing roots in a compost heap to form hot spots, but a heap would, given the right conditions, create its own hot spots of activity.

Three ways to compost

Traditionally, more interest and emphasis have been placed on the end product and uses of compost than on the composting process itself. There are three types of man-made compost heaps: anaerobic (cold, without oxygen), aerobic (hot, requires oxygen) and vermicompost heaps (with worms). A vermicompost heap requires oxygen but does not get hot. Many people compost using the anaerobic method (by default); some more skilled composters use the aerobic method; but the enlightened amongst us harness the power of worms to great advantage.

Dominant microbes

Compost heaps are dominated by millions of microbes, so-called 'microherds'. Thermophilic microbes are dominant in aerobic heaps (their activity generates the heat) and anaerobic ones (that thrive in the absence of oxygen) are predominant in anaerobic heaps. A different assortment of microbes will be predominant in vermicompost heaps. However, no two compost heaps will have the same mix of microbes - it will depend on the materials in the heap and all the various environmental factors.

Vermicomposting

The idea of vermicomposting is to use a large population of selected earthworms to eat your organic waste. By protecting them from predators and optimising their environmental conditions you can maximise their efficiency. This allows them to rapidly increase in numbers and process the available waste.

The worm advantage

In the wild, some worms ingest dead plant material. On damp evenings you can witness worms pulling dead leaves lying on the lawn down into their burrows. The worms eat the dead plant material, which will be teeming with bacteria and fungi. The worms will also ingest particles of soil, sand, silt, or clay, and these will also contain living organisms such as protozoa, nematodes, minute creatures or their eggs.

All these things are mixed up in the gut of the worm, where conditions are perfect for the bacteria and fungi to speed up their processes, and decompose the organic matter more quickly. In this way worms are similar to herbivores, such as cows. Cows eat the grass but it is the microbes in their stomachs that actually digest it, breaking it down into forms that the cow can use. The microbes take their 'tax' of nutrition. As with the cow, the worm takes nutrients from the material and poops out the rest. The material that comes out of a worm is very different to what goes in, with greatly altered levels of organic matter, bacteria, fungi, protozoa and nematodes.

Scientists have not been able to demonstrate that worms themselves are capable of producing the enzymes (such as cellulase) to break down the various constituents of the dead plant material. Do the worms consume the decaying plant material because it contains the microbes that can do the job, or because the microbes have already made it edible to them? Scientists do not exactly know. They do know that the growth of bacteria, microbes and fungi is enhanced by being in the worm's gut, and remains so when expelled at the back end of the worm.

Vermicompost is rather special

As well as the enhanced microbial activity in worm casts, the passage through an earthworm's gut alters the organic material. The 'product' of all the worm and microbial activity is rather like peat, with good structure and high nutrient content. It allows good aeration for root growth and has good water-holding capacity.

Compared to garden compost produced by the alternative methods, vermicompost is far superior for healthy plant growth. It is 'biologically enhanced' by the worm activity - a rich source of beneficial microbial life, which improves soil structure and texture, makes nutrients available to plants and improves disease resistance. When you add vermicompost to your garden, you are inoculating it with this rich and diverse microbial life.

Plant nutrients present in vermicompost are readily available to the plants, and it compares very favourably with most commercial plant growth media, like potting composts, which have inorganic chemical nutrients added to them. In addition, vermicompost contains natural growth enhancers, by virtue of the microbial activity.

Vermicompost is perfect for use in the garden or for houseplants, and an extremely useful liquid feed can be made from it. It is a quality product and when produced on a commercial scale it demands high prices. Compared to most other fertilisers and soil conditioners, either chemical or organic, it is very expensive. As a vermicomposter you will have a free and ready supply. The ultimate in home waste management, using your organic waste to enhance the health of the soil in your garden.

There will probably be some worms in a non-vermicompost bin, even quite a lot of worms, but in the main the compost will have been broken down by microbes rather than in the gut of a worm.

Benefits of vermicompost at a glance:

- Full of beneficial soil micro-organisms
- Very high humus content
- Contains slow-release natural fertilisers
- Nutrients available in form that plants can readily use
- Absorbs 2-3 times its own weight in water
- Holds water well so less watering required
- Helps to bind together soil particles
- Enhances disease resistance
- Encourages healthy/strong root system
- Produces strong healthy plants/crops
- Improves crop yields
- Liquid fertiliser can be produced
- It's free!

Vermicomposting vs composting

Not only does vermicomposting produce a fantastic 'biologically enhanced' product, but it also has several advantages over garden composting, either anaerobic or aerobic. Food waste can be fed to the worms all year round, adding small amounts on a regular basis. There are some waste items that people are reluctant to put on their compost heap, which can be fed to worms. A worm bin can be sited indoors or outdoors and requires minimal space. It is easy to manage and does not require turning like a compost heap – the worms do the turning for you, and by their tunnelling actions aerate it as well. And a bonus for fishermen – it supplies a steady and immediate supply of bait!

Responsible Waste Management

It seems obvious today that recycling should be the absolute priority for waste management. Vermicomposting has been recognised world wide as an excellent process for more sustainable (or environment-friendly) waste management. There are many examples worldwide of large-scale waste materials being successfully vermicomposted:

- Paper and cardboard
- Vegetation from parks and gardens
- Human sewage ('biosolids')
- Animal manure
- Household waste
- Food waste from factories, distributors and the catering trade

Worms can deal with all of these wastes and recycle them into a useful resource.

Household waste

On average about 38% (sometimes more) of a household bin is organic material. This could be turned into a valuable commodity if we fed it to worms. Domestic waste as collected is unpromising to feed directly to worms, because many inorganic materials such as plastics, glass and metals are mixed in with the organic waste. Such materials will not be broken down by the process and are potentially toxic to the worms. All of us could process this organic waste material ourselves by vermicomposting it. As well as the waste management benefits, we could all then use the products of this process to our advantage. Food waste, which many people are reluctant to throw onto the compost heap, can be fed to composting worms. With more and more waste being created, less and less space available to accommodate it, it is time for us all, politicians, waste managers, local authorities, gardeners, hotel owners and school children, to wake up to the potential of worms. We must begin to take responsibility for the waste we produce, and vermicompost it at home.

Using a wormery: Advantages at a glance

- Best practicable environmental option for your food waste
- Most efficient means of indoor composting
- Can be done all year round
- Requires minimal space
- Easy to use and manage
- Less time-consuming than garden composting
- Composts small amounts on a regular basis
- Clean, safe and odour-free
- Can be used indoors or outdoors
- With some wormeries, option to observe process and worms
- Produces FREE fertiliser for garden use
- Steady and immediate supply of fishing bait
- Emotionally rewarding
- An ideal educational tool for schools.

Increasing world-wide interest in worms and vermicomposting

The process of vermiculture and value of vermicompost, together with the amount of available information about earthworms has increased hugely over the past few decades. A quick search of scientific literature reveals a multitude of references about vermicompost and the World Wide Web makes available to Internet users a huge amount of scientific and anecdotal information.

There are thousands of people in Britain who vermicompost their kitchen and garden waste, but there is not such a range of worm bins available commercially in Britain as there are in the USA or Australia. In the USA, for example, thousands of schools use worms to compost their waste, and worms were used to vermicompost the huge amounts of waste generated by the Sydney Olympics in 2000.

Domestic vermicomposting in the UK

What we need here in Britain, and have lacked up until now, is a combination of people from different disciplines to work together. Once we can get people from scientific research and engineering; people with practical vermicomposting expertise; environmental educationalists and those with marketing experience to work together then things will really move forward. Thankfully, this is beginning to happen and things are at last beginning to move in a positive direction.

There is only a small but growing handful of companies here that sell wormeries, and until recently the models available were quite frankly poorly designed and needed constant management. Things are starting to change - vastly improved wormeries are now available, with advice and information easier to access.

The Biology of Earthworms 2

"Worms seem to be the great promoters of vegetation, which would proceed but lamely without them..."
The Rev. Gilbert White of Selborne, 1777

Earthworms are among the most ancient animals on earth. Sadly, as they have soft bodies, which do not lend themselves to being fossilised, worms are not well represented in the fossil records. These records are dominated by animals with hardened body parts, yet worms have been recorded as far back as 650 million years ago.

Worms are such successful animals that they constitute a significant proportion of the mass of the world's soils. Their numbers can be as high as 100s, even 1000s, in just one square metre of soil.

Most of us are unaware how important worms are. They are vital for the development and functioning of soils, as we know them. Perhaps it is not surprising that over the centuries interest in worms has been considerable. The ancient Greeks considered them to be the 'intestines of the soil' and Cleopatra decreed that the earthworm be revered. Her subjects had to protect it as a sacred animal, and Egyptian farmers were forbidden to 'trouble the worm or remove it from the land' as wisely it was thought it would affect the fertility of their soil.

There are already thousands of known earthworm species and yet earthworm taxonomist Sam James keeps on finding and naming more. No doubt there many more worms yet to be found.

The type of worms we are interested in are 'humus formers', which belong to a family called Lumbricidae. These worms can be used to convert our food and other waste into vermicompost. In Britain, it is generally accepted that there are about 27 species of Lumbricidae, and a few more in the USA, many of which were introduced from Europe.

What is a worm?

At first glance an earthworm appears to be a very simple animal. It is apparently blind and deaf; without legs or feelers; it is difficult to distinguish front from back and top from bottom; it has no obvious strength in its body and little to protect it from the elements or predators. However, it is a hugely successful creature and, as you will see, there is more to it than meets the eye.

Basic design

Basically, earthworms are muscular cylinders filled with fluid. From the worm's mouth right through to its rear, is a long thin tube, which contains its food processing systems, such as its crop, gizzard, gut and intestines. Between the inside and outside walls is a cavity called the coelom: this is segmented and filled with body fluids collectively known as coelomic fluid. At one end of the worm is an opening that serves as a mouth, and at the other end is the anus. A little known fact about the worm is that it has five 'hearts'.

Breathing

Worms don't have lungs, they breathe through their skin. They can survive in water for a time, if it contains plenty of oxygen. When it rains worms can often be seen on the ground, and some people think that they have left their burrows to avoid drowning. We do not know why worms come to the surface in rain, but it is not because they would drown in the ground. It may be that they are looking for another worm with which to mate or it is easier for them to travel across wet ground.

Senses

As well as the gut, a nerve cord runs the length of the body through all the segments, connecting various sensory organs on the outside with the tiny, primitive brain. The organs are specialised to a variety of stimuli, allowing worms to sense when they are being touched, feel the vibrations of certain sounds, sense certain chemicals in their surroundings, and some are sensitive to light.

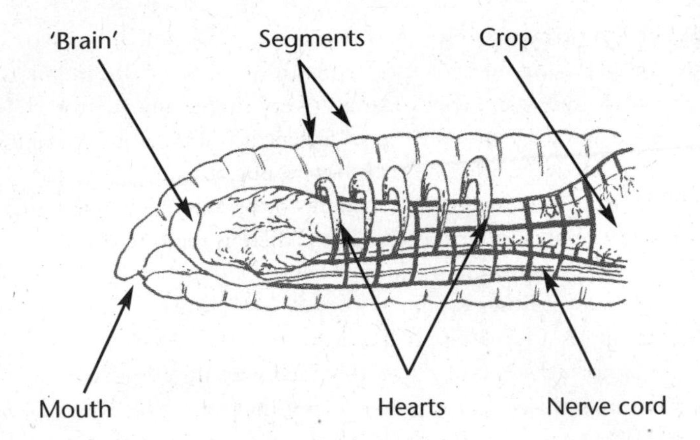

Moisture

Worms obtain their water from the food they consume as well as from water-filled pores in their environment or from surface films. Their own water content is greatly influenced by the water content of their environment, and is critical to their performance. Worms kept in well-aerated water are fully hydrated, with water making up about 85 % of their fresh weight. Most of this is in their blood and coelomic fluid. Under field or laboratory conditions, this usually falls to about 65-70%.

Mechanisms for water conservation are poorly developed in worms. They lose water though their skin as well as openings from their bodies and a lot of water is lost in urine, so they are dependent on a moist environment. Many worms enter a dormant phase if they become dehydrated. However, brandling worms can lose over 60% of their body weight, which is over 70% of their water content. This is like a 65 kg (ten stone) human being dehydrated to 25 kg (four stone). However, unlike the human, the worm will recover when placed in a moist environment. Obviously this is not ideal and doesn't make for particularly happy worms.

Some species of earthworm can survive in waterlogged conditions, but a waterlogged worm bin is not suitable for your composting worms. They can survive these conditions for a very short period, by drawing upon a small stored energy source. In nature, they would move away from this unsuitable habitat, However if enclosed in a small wormery there may be no 'safe' areas to move to, so they will try to escape or congregate in an area away from the hostile environment, such as the lid to the worm bin.

Mucus

Worms produce mucus, secreted by glands in the skin. This helps stop the worm drying out, just like a slug, but they can only make it if they are in a moist environment. Mucus is very important to worms in a number of ways: it allows them to breathe (which they do through their skin), it acts as a lubricant helping them to move through the soil, it forms a protective layer along their skin, and binds soil particles together in their burrows.

Movement

Worms have several features that in combination allow them to move. There is a flap of skin covering the mouth opening (called the prostomium) so they can move through soil without having to eat continuously.

Each segment of the earthworm's body acts as a separate hydraulic system and through co-ordination of the segments worms are surprisingly rapid movers. Movement is the result of waves of alternate contraction and relaxation of the opposing muscle layers.

All along their bodies worms have tiny hairs or bristles (setae), which allow them to gain some purchase as they push or pull themselves through soil (or are pulled by a blackbird!). We don't usually think of worms as hairy, quite the reverse in fact, but if you put an earthworm on a piece of paper, you can hear these bristles as it moves around, similar to the sound of a man stroking his unshaven chin!

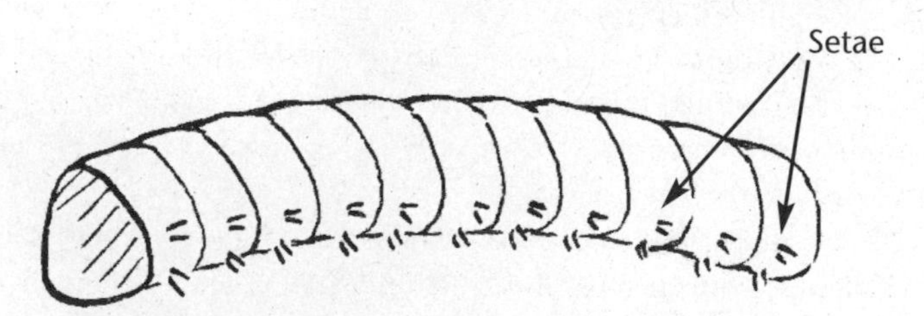

The final key to the movement of the worm is its mucus that acts as a lubricant.

Diet

As worms do not have a stomach, they select food that is already being decomposed by microbes, then their gut flora and fauna continue with the decomposition and breakdown of the food material. Only a small proportion of a composting worm's nutrients come from the decomposing organic matter that it eats - most of it is from micro-organisms and in particular fungi. Some of these microbes survive the worm's digestive system, and in fact flourish in the worm's gut before being expelled at the other end. The organic material that the worm eats is, therefore, used by the worm as a carrier for the digesting microbes rather than being the sole source of nutrition. An earthworm is essentially a microbe predator.

Worms do have a gizzard, like a bird, containing particles of soil or sand that grind their food into fine particles.

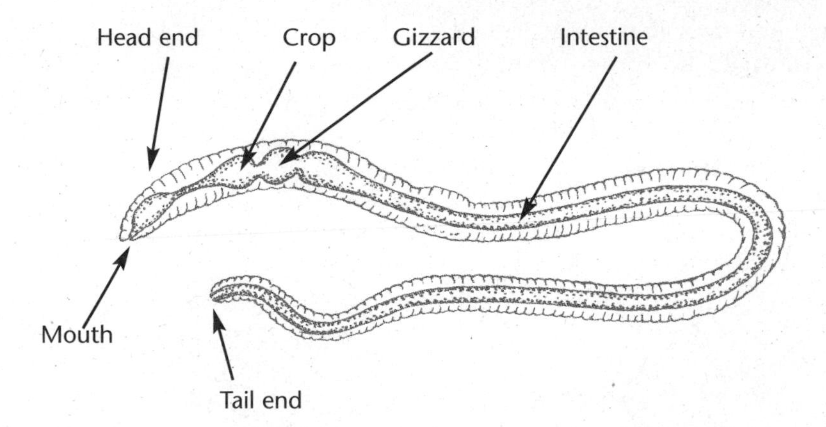

Temperature

Temperature, of course, affects all stages of worm life. Laboratory studies have provided much information about the temperature ranges that worms can tolerate. Many of the common earthworms will die at just below freezing point, although many survive by migrating to deeper soil where it is warmer. Most worm species die as a result of prolonged exposure to temperatures between 30-40°C. They are often most productive at around 20°C, but different species have different optimum temperatures.

A happy, hungry sexually active worm needs a warm, damp environment. Unfortunately, cold and wet usually go together in the UK.

pH

pH is a measure of how acidic or alkaline a soil is. It is shown on a scale of 1 to 14, 1 being highly acidic, 7 neutral and 14 highly alkaline. The humble earthworm has an incredible pH tolerance and is found naturally in environments with a pH anywhere between 4 and 9. The pH in organic matter shifts as it decays, and composting worms have evolved to cope with this. Their optimum pH is a mildly acidic 5, and not a neutral 7.

Age

Adult worms may live for many years in the absence of predators or extreme environmental conditions, although in the wild they probably only survive for a matter of months as an adult.

Reproduction

Worms are hermaphroditic (which means they all have both male and female sexual organs), but most still need to mate with another individual to produce offspring. Members of some species can reproduce without being fertilised by another worm.

Worms find appropriate mates by recognising glandular secretions.

Consenting worms lie side by side with their heads in opposite directions and direct contact is made between their opposing sex organs. Sperm produced by the male organs (called the 'male pore') is exchanged and stored in sperm sacs.

After mating, a worm secretes mucus from its clitellum or saddle (the swollen section visible on adult worms). The mucus hardens and the worm backs out of it, leaving its eggs and sperm inside. The ends of the mucus tube close around the eggs, forming a resistant cocoon. Cocoons are lemon shaped and may hold up to 20 eggs, but usually only one or two will survive. Cocoons can withstand desiccation, damage and cold conditions that would kill adult worms, and are the main survival strategy for some species.

Habitat

Earthworms live in the soil, but not just anywhere. Different species live at specific levels, reflecting their burrowing habits and feeding patterns. They have been classified into three major ecological groups:

Surface feeders (epigeics) eat freshly decaying plant or animal residues and consume little or no soil. Living in or near surface plant litter, they have adapted to the highly variable moisture and temperature conditions at the soil surface. They are unlikely to survive in the low organic matter environment of soil. They are typically found in deciduous woodlands, compost and manure heaps. They tend to be small with high rates of reproduction, producing large numbers of cocoons. For worms, they are rapid movers, but being a favourite food of birds and mammals they tend to be short-lived.

Shallow-living worms (endogeics) live in horizontal burrows in the upper soil levels, usually about 10 -15 cm down. They feed primarily on soil and associated organic matter in the root zone. They do not have permanent burrows, and their temporary channels become filled with cast material as they move through the soil. The worms are medium sized, of intermediate longevity and suffer low predation rates.

Other species (anecics) live deep in the soil in permanent or semi-permanent burrows, emerging onto the surface at night to forage for freshly decaying organic matter. They drag decaying organic matter down into their burrows in the soil. Burrows are semi-permanent and may be over 3m deep. The worms may leave plugs of organic matter, small stones or casts, to 'cap' the mouth of their burrows. These worms are usually large and sluggish, relatively long-lived, but produce few cocoons and have high rates of predation when at the surface.

In a standard pasture, the proportions of these different types of worm would be about 40% surface dwellers, 10% shallow working species, and 50% deep burrowers.

Which worm do you want in your wormery?

If we went out and found worms in the garden or the park and put them in our wormery, they would probably not do very well. This is because it is highly unlikely that they would be the right species. For a start we want certain humus formers, not humus eaters. We need to carefully select worms that are capable of processing large amounts of organic waste. Ideally they would also be able to reproduce quickly, tolerate disturbance and be native to Britain so they can thrive in our climate. Also, they would have to thrive in the confines of a worm bin - a tall order for the vast majority of worms.

Where would we find such a worm? The surface dwellers, that eat fresh organic matter, would be a good place to start. Some of these worms can be found in manure, dung and compost heaps throughout the UK, and perhaps some of these species would make the ideal addition to our wormery.

Below, I have briefly described each of them, as well as Lumbricus terrestris, the common garden earthworm that most of us are familiar with. Lumbricus terrestris is a humus former but not a composting worm. Although some of these species will grow satisfactorily and survive well in a wide range of organic wastes, research has shown that some are more prolific and grow faster than others.

Dendrodrilus rubidus

This small earthworm ranges in colour from dark red to pink on the upper surface, being paler below. It often has a conspicuous yellow tip to its tail, so anglers call it 'gilt tail'. It is common in the surface layer of decomposing leaves in woodland and can sometimes be seen underneath the bark of decaying logs. It can be found under dung in pastures and is often very abundant in compost and manure heaps.

Although it would make a decent composting worm, it has a fairly slow rate of reproduction so is outperformed by other worms.

Eisenia fetida

This moderate-sized earthworm has a very distinctive striped appearance, referred to by the common names 'brandling' and 'tiger worm'. The pink to purplish-red pigmentation occurs in bands separated by unpigmented areas that have a yellowish hue. When the worm is irritated it exudes a musty (foetid, hence the Latin name) yellow fluid through small pores on the upper surface. This seems to be a defence against predators, and as is often the case with animals that taste bad, the stripes are a warning.

The brandling is sometimes found in deep woodland leaf mould and it can be extremely abundant in manure and compost heaps. It naturally colonises a huge range of different organic wastes, with varying ranges of moisture content. Brandling worms grow and reproduce extremely quickly, they are tough, readily handled, and tolerate a wide range of temperatures. They can live communally in very large numbers and are the choice of commercial vermicomposters and waste processors throughout temperate regions of the world. Being native to the UK they are ideally suited to vermicomposting here.

Eisenia andrei

The only visible difference between E. andrei and E. fetida is the stripes - E. andrei is uniformly reddish. It is sometimes called a 'red tiger worm' and until recently they were thought to be one species with colour variation. Apart from the colour they are very alike. Many commercial worm farmers unknowingly grow both species, mixed together and do not separate them - which would be very difficult and time-consuming, even if they could positively distinguish them! As they are so similar I will not distinguish between them and refer to them jointly as E. fetida.

Eisenia hortensis (formerly Dendrobaena veneta)

This earthworm is commonly called the European Nightcrawler. It has a pinkish colour at the front end of the body but is mainly greyish when, as is usually the case, the gut is full of soil. The tip of the tail is cream or pale yellow. It is larger than the brandling worm, E. fetida, with a much lower reproductive rate (about 10 times less), growing rate, maturity rate and hatchling rate from cocoons. It is found in deep woodland litter and in garden soils that are rich in organic matter, including under compost heaps. Its preference is for damp soils, more so than E. fetida.

So E. hortensis can do very well in worm bins if you keep it quite wet, and as it is a popular bait amongst anglers it makes a good choice of composting worm if you are a fisherman. It will eat your garden and food waste and become fat for the hook.

However, I would always have to recommend E. fetida for non-anglers, as it really has many advantages as a composting worm. The much faster rate of reproduction means that it will fill your bin much more quickly, and you can begin to harvest compost earlier. E. hortensis has a narrower temperature tolerance and becomes more stressed outside of its optimum temperatures than does E. fetida. Trials using E. hortensis and E. fetida together in worm bins found that E. hortensis tended to remain near the bottom of the worm bin, where it was wetter and E. fetida remained nearer the top where it was drier. E. fetida dominated the worm bins, by virtue of its prolific breeding capacity. In wormeries that use the continuous flow system, as E. hortensis prefers the damper conditions near the bottom of wormeries, they tend to fall out. So the solution for anglers is a mixture of both worm species.

E. hortensis is used for large scale in the specialised vermicomposting of potato and paper pulp waste, which are wet, and in these situations it outperforms E. fetida.

Eisenia veneta

This earthworm is like a large version of E. fetida, with its stripy appearance and the same habit of exuding yellow fluid onto its skin when irritated. It is also found in the same kinds of habitat.

E. veneta reproduces more slowly than E. fetida, but being a large worm is popular with anglers for bait.

Eudrilus eugeniae.

Commonly called the African Nightcrawler, this is a great composting worm, with a rapid rate of reproduction and food processing. However, as it comes from Africa it likes it hotter than it usually is in the UK, and is difficult to keep in captivity. It is widely used in the USA.

Lumbricus rubellus

This moderate sized earthworm is purplish red in colour on the upper surface and pale below. It has been called a 'red worm' or 'marsh worm'. It is found in a very wide range of soils, being especially abundant where the soil is organically rich, such as under dung-pats in pastures. These worms are migratory, moving into fields that are freshly manured and moving on once it has been spent.

Although found in manure and composting heaps, indeed sold as a worm for vermicomposting purposes, L. rubellus has not actually been identified as a good worm to use for vermicomposting purposes. It is a soil dweller that happens to like a good organic matter content in the top few inches of soil.

Lumbricus terrestris

The largest British earthworm (up to about 300 mm) is purplish-red above, the pigmentation often being largely present at the front end of the body, on the upper surface. The end of the tail is often distinctly flat. It is found in a wide range of soils where it burrows deeply, down to over a metre. The hind end of the body usually remains in the burrow, anchoring the worm so that if danger threatens it can rapidly retreat to safety. You will have seen this worm on your lawn, especially on damp evenings, when it pokes its head out of its burrow in search of food or a partner.

This species has a variety of common names including lob, dew worm and night crawler.

A note on names

I have included some of the more common names for these worms, but they do vary between regions, and are open to extreme confusion. Do you want a red worm, tiger worm, blue nose, Nightcrawler, gilt tail, red wiggler, manure worm, Belgian nightcrawler, dendro, Canadian nightcrawler, brandling or red tiger worm? That is why it is easier to use the scientific names for worms, which are used worldwide and should be standard everywhere.

However, even the salesmen who ought to know can get it wrong. L. rubellus has been sold for many years as a composting worm, but in fact, when professional worm taxonomists identified them, they were actually our champion composter, E. fetida. To be fair, worm identification is a great challenge. The size and location of both internal and external structures on the worm body, which in the main need to be viewed under magnification, are critical to determine which species you are looking at.

What's living in the bin with the worms? 3

"Earthworms are soil builders, everything else – plant, animal, man and bacteria are food for earthworms whose function is to mix living matter with mineral particles and send them forth on their round once again."

From 'Harnessing the Earthworm' by Dr. Thomas J. Barrett 1947

Microscopic organisms - the 'microherds'

Inside a worm bin is a very complex food web, with most of the players out of sight to us humans. The organisms that predominate vary depending on the state of the materials in the bin, fluctuating as the conditions change through the process of decomposition. The function and success of each organism in the web depends on what food is available at a particular time.

A compost heap can contain millions of different varieties of soil-dwelling microbes, mostly unknown to science. I can only mention here a small selection of the more important ones. This will hopefully give you some idea as to the complexity of life that exists in compost heaps and worm bins.

Actinomycetes (ray fungus)

These organisms can sometimes be seen as greyish webs stretching through soil or compost. These long branched filaments were thought to be fungi for many years, but in fact are fungi-like bacteria They are primary decomposers and are especially effective at breaking down raw plant tissues (cellulose, chitin, and lignin). They can decompose tough materials like bark, newspaper and woody stems, softening them for other organisms. They are said to be responsible for giving us the typical earthy smell of compost. Actinomycetes are of great interest to many scientific disciplines, not least medical science - we can thank them for the antibiotic streptomycin.

Bacteria

These are perhaps the smallest and most numerous living organisms in compost heaps - there will be millions of them. We tend to associate bacteria with dirt and germs, but in fact they are vital to life on earth, and particularly important in a worm bin. They help decompose plant and animal material releasing nutrients, and they are food for many bigger creatures, including the worms.

Fungi and Mycorrhizal fungi

There are millions of fungal spores floating around in the air and they will happily make a home and start to work for you in your worm bin. The most beneficial fungi are known as saprophytes, which feed upon and break down organic matter, producing humus. Sometimes you may see fungi on the top of your worm bin, in the form of a white powdery layer or thread-like hairs.

In evolutionary terms, while most plants specialised in absorbing sunlight, fungi focused on absorbing nutrients. They are unusual in that they can break down lignin and cellulose in wood and other plant materials - very useful in a compost heap. They also provide a nutritious feast for the other creatures in the worm bin, including worms. Mycorrhizal fungi are immensely important to agriculture and gardeners as they form a beneficial symbiotic relationship with plants. Indeed perhaps 90% of the world's land plant communities have formed such relationships. These fungi are found on plant roots and in some research it has been shown that worms can greatly enhance the number of the mycorrhizal fungal spores. This will in turn greatly benefit your plants when grown using vermicompost.

Nematodes (also known as eelworms, roundworms)

It is very likely that these microscopic worms will be in your worm bin.

Although nematodes may not contribute directly to the decomposition process in worm bins, they are an important link in the food chain. Different nematodes feed on different foods, including fungi, bacteria, microbes, other nematodes, rotifers, protozoa, slugs, beetles, mites and the liquids that come off rotting organic matter. And they themselves may fall victim to predacious mites or springtails.

Protozoa

Although some protozoa feed on others, mostly they feed upon bacteria. This might at first glance appear to be a bad thing for the system, as bacteria are so important to the decomposition process. But by eating the bacteria, the protozoa may add to the nutrients available to other organisms, and protozoa make an essential part of the diet of some species of earthworm, especially our friend the composting worm, E. fetida.

Many factors determine the numbers of protozoa in worm bins, such as temperature and available food. They are capable of encysting when conditions become unfavourable, becoming active again when conditions improve - a kind of hibernation.

Rotifers (wheel-bearer)

So called because they appear to whirl like a wheel when they propel themselves forward. They can be found wherever there is moisture, from puddles and ponds to worm bins. They feed by capturing, grasping or filtering their prey or food. Some are known as vortex feeders (like a whirlpool), creating currents in the water in soil particles by moving tiny hairs rapidly, and directing particles of organic matter and tiny soil micro-organisms to their mouths.

Rotifers have an amazing ability to survive long periods without water, shrivelling up (encysting), and then when the right conditions come along, bingo, they wake up and carry on where they left off (excyst).

Macroscopic residents (or things we can see)

Once the microbes have started to decompose the material in your compost heap or worm bin, the big boys move in! If a worm bin is indoors, many of the larger creatures that would normally find their way into an outdoor heap, such as ground beetles, centipedes and ants, may never enter the bin. If you collect material from the outdoor compost heap, you may transfer such creatures into the worm bin unwittingly.

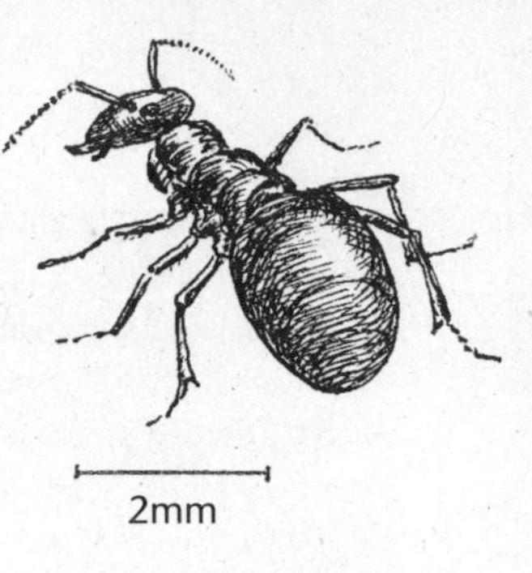

Ants

Black garden ants are familiar species and need no introduction or description. The ants may be visiting for food or perhaps have made their nest in the bin. In the UK, black garden ants do not make a meal of the worms themselves. They are harmless in a worm bin, although to worm breeders they may cause a problem when selling worms if ants or their eggs are sold as an added extra! However, if you have ants in your worm bin it indicates that it is too dry - gently sprinkle with water to deter them. If you find ants and their eggs in the bin, disturbing the area with a fork regularly for a few days, together with watering, would give them a huge hint to move away.

Red wood ants

If you have the red wood ant (Formica rufa) in your bin it may be a different matter. These ants are common in many parts of mainland England and Wales but predominately in the south. They build their large nests from decaying vegetation, and obtain 80% of their food from 'milking' carbohydrates from greenflies and scale insects found on plants. They require the remaining 20% as protein, which they get from animal prey such as insects, caterpillars and grubs etc. Adults chew the prey and regurgitate it for the developing larvae.

Some species of worm have been found to positively thrive in the uppermost layers of the nests of red ants. Conditions are perfect for worms, always moist, ample supplies of decaying vegetation, warm and safe as ants eat any potential worm predators. The coating of mucus on the worms deters the ants from eating them. They may also have the ant colony smell which would protect them from the ants. Worms in their turn benefit the ants by eating the fungi and moulds that would otherwise rot the nest away.

However, most worms are not lucky enough to be the protected guardians of the red wood ant nest. For the outsiders it is another matter. Worms have been seen cut in pieces and carried to the nest as food. So if you live near a forest or wood and have a nest of red wood ants nearby, they may find your worms and see them as a tasty snack through the spring and summer months. As well as your worms they would eat other beneficial insects inside a wormery.

As with black ants, keeping the wormery moist will deter red ones. A few sudden downpours of water will be most unpleasant for them. Some wormeries have metal or plastic legs that can be placed inside small tin cans containing a little oil, forming a moat that the ants will not cross. If nothing else works, borax powder mixed 50/50 with caster sugar and placed near an identified ant run will also see off the ants.

Bathroom fly *(also known as a moth fly, sewerfly or drainfly)*

As the common name implies, these tiny flies are usually found in bathrooms and lavatories. They are about 4 mm long with a somewhat fuzzy appearance. The wings appear too large for such a tiny fly. Like moths, they hold their wings roof-like over their bodies when at rest. They do not fly particularly well, making irregular short-distance flights of just a few feet, or they simply walk around. Their larvae, about 5 mm long, are flattened, with a distinctive head and small suction discs along their underside. They are legless and live in the slime around taps and plug-holes.

In wormeries they are considered beneficial to the decomposing process, as the larvae help to break down organic matter. If present in a wormery in large numbers the flies can leave in a cloud when the lid is removed. Outdoors this is not usually a problem. In large numbers though, indoors, they can be a nuisance (*see Chapter 7, Frequently Asked Questions, on controlling fly populations*).

Centipedes

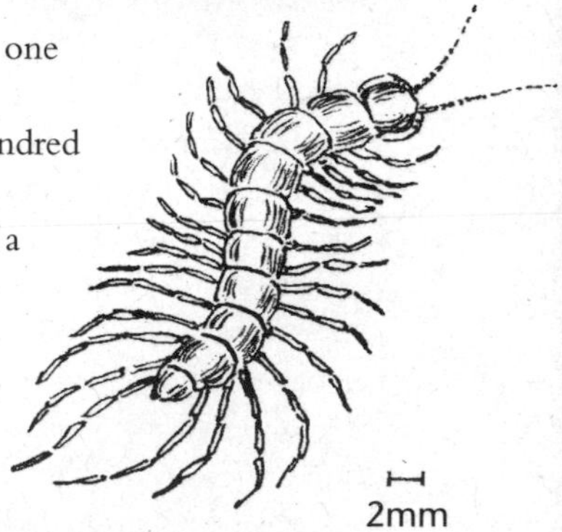

Literally 'one hundred feet', but what we mean is one hundred pairs of legs! Of course this is not quite accurate, several species having more than one hundred pairs! Centipedes can be distinguished from millipedes because each definite body segment of a centipede has one pair of legs.

Centipedes are generally carnivorous, equipped with powerful front claws (which are in fact modified front legs). These claws can inject lethal venom into any small insect prey (don't worry, British ones can't harm humans!) At certain times of the year leaf litter may form part of their diet, perhaps in the winter when prey is hard to find. They are nocturnal, so are only usually seen in the day if disturbed.

2mm

Broadly speaking there are two types of centipede - fast running surface dwellers and longer, thinner ones that live below the soil surface, sometimes referred to as 'snake centipedes'. Either type may enter worm bins, and a few centipedes will not cause much damage by way of eating your worms and other beneficial organisms. If you want rid of them, simply pick them up and put them into the garden where they themselves are beneficial, eating slugs and other small nasties.

Collembola (spring tails)

Collembola are primitive wingless insects, mostly living in soil and rotting vegetation. They are one of the most abundant and widespread groups of soil arthropods, with species found in deserts, caves, snowfields and forests. Under the right conditions populations can be very large, up to 100,000 per cubic meter of soil. Their common name, springtail, derives from the forked tail that enables them to jump quite considerable distances, 7-10 cm. This is some feat for an insect that is a few millimetres long at most! When not in use the tail is folded forwards into a groove on the underside of the abdomen. They move by crawling or jumping, followed by periods of rest.

2mm

Springtails eat a wide range of foodstuffs, depending on where they live. Some can chew their food and others suck up juices. They are extremely beneficial in worm bins, speeding the process of decay. Sometimes they can cause alarm when discovered in large numbers, but they are completely harmless to humans.

Fruit/vinegar flies

Fruit flies can be recognised by their habit - hovering around decaying fruit and vegetables. They are especially prevalent in the summer and autumn. They are small, about 4 mm long, brown, and have red eyes (look closely!). The larvae feed principally on yeast in the fluids produced as fruit decomposes, and crawl to a drier area to pupate. Under favourable conditions, the whole life cycle can be completed in a little more than a week. The vast majority of fruit flies are introduced to worm bins as eggs attached to scraps of fruit and vegetables. The damp environment in the bin is ideal for the eggs to hatch. They are beneficial in a worm bin, but that is no consolation to you as you fish out another drunken fruit fly from your well-earned glass of wine! They may be a nuisance by their numbers, especially indoors - see *Chapter 7, Frequently Asked Questions*, for how to manage your bin to keep fly populations under control.

Millipedes

Millipedes inhabit a wide range of habitats around the world. Often confused

with centipedes, they have two pairs of legs per segment, unlike the centipede that has one. They are less active than centipedes and although they have more legs they move more slowly. Juveniles resemble insects, with only six legs, so they have a lot of legs to grow to become a fully fledged millipede!

Millipedes eat decaying material, wood and leaves by preference, though some will eat carrion and fungi. Their main contribution to the worm bin is the mechanical breakdown of decaying matter, making it more susceptible to microbial action. Some species of millipede are obligate coprophagists - they must eat their own faeces to remain healthy. This suggests a close relationship with their gut bacteria!

Mites

Mites, with their round bodies and eight legs, are often mistaken for spiders. As a group they are very diverse, but one will be familiar to gardeners - the red spider mite (Tetranychus urticae). This notorious pest is often found in greenhouses on your tomatoes. Most mites thrive in undisturbed moist organic matter. They are beneficial to the composting process, feeding on fungi, fungal spores, dung and detritus. Other species consume nematodes, collembola, other mites and small insects. Large fluctuations in numbers and species of mites present occur naturally through the year as environmental conditions in the wormery vary.

Fungus gnats and sciarid flies

There are many, many hundreds of species world wide within this large family. The larger species that you may see flying around your wormery are often called fungus gnats. Their larvae feed on decaying organic matter. They look rather like mosquitoes in size and shape but they are, in fact, completely harmless to humans, lacking the piercing mouth parts and bloodsucking habits of mosquitoes. They are not likely to be a problem in worm bins indoors or outdoors.

Sciarid flies are small, black midge-like flies about 3 mm long. They are quite dainty with long antennae and very long legs. They fly around with a humped back and dangly appearance. They can often be seen indoors, running fast across plants or the compost around houseplants. The larvae are transparent and grow up to 5 mm long. They feed for up to two weeks on eating fungus and other organic matter. Once fully fed the larvae pupate in the compost and hatch some days later. Given favourable conditions the whole life cycle can be completed in about a month, with a continual breeding cycle all year round.

Both the larger fungus gnats and the sciarid flies are not a problem in a wormery or compost heap. However in the absence of decaying organic matter their larvae may feed on the roots of indoor house plants.

Potworms (Enchytraeids)

Potworms are small white earthworms, sometimes mistaken for nematodes. They range in size from 1 mm to as long as 50 mm. They are a prized food for tropical fish, native birds, frogs, beetles, predatory mites, etc. In the wild they live in permanently damp soils with a high content of organic matter, and they will be happy in your vermicomposting system! Not only are they harmless to you and your worms, they are beneficial residents in your wormery. They feed on decomposing organic matter, such as plant fragments, fungus and silica grains if available, adding to the overall decomposition. Their manure will add to the rich diversity and community of microbial life in the system, as well as providing food for the larger worms. They are thought to control some nematode pests by eating them.

Pseudoscorpions or false scorpions

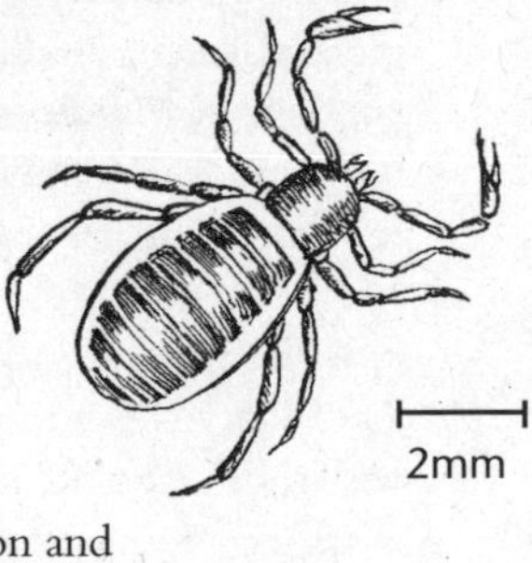

Don't be alarmed if you see these tiny scorpion-like creatures in your worm bin! They resemble true scorpions in everything except their size (the largest species in Britain is only about 4 mm long), and lack the tail and sting. They are quite harmless to humans and animals. Phew!

2mm

Pseudoscorpions are found in moist, decaying vegetation and are unlikely to become abundant. They feed upon other litter-dwelling creatures such as mites, collembola, small insects and smaller pseudoscorpions. Tactile hairs alert them to their prey, which they ambush, seizing it with their huge claws. They inject the prey with a poison and suck the juices out, ready to fend off any attackers or would-be food robbers with their claws! They are able to go for long periods without food, but when prey is available can feed continuously for many hours. Like true scorpions, the males perform a ritual display before mating. Females carry the eggs around in a pouch and when the young are born they may hitch a ride for some days. Adults can live for up to 5 years, over-wintering in silken chambers that they secrete. They can walk forward or backward and are fascinating to watch.

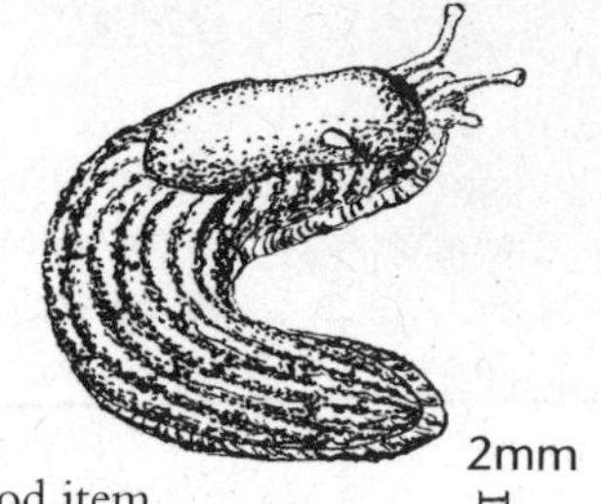

Slugs and snails

These need no introduction at all! Most slugs species have no shells, all snail species do. They have voracious appetites, as all gardeners know, using a file-like structure in their mouthparts to scrape or rasp away the surface of their chosen food item. Contrary to popular belief, some slugs and snails can be beneficial to the decomposition of organic matter. Many slugs (and to a lesser extent snails, restricted as they are with their large shells) feed on surface rotting vegetation and fungi and then move down into the sub-surface layers of the soil, incorporating organic material into the mineral structure of the soil. Some species of slugs and snails produce enzymes in their gut capable of digesting cellulose, making it readily available for the soil microbes to consume. Mucus, produced by slugs in great quantities, may serve to promote the development of water-stable soil particles, thereby contributing to a good soil structure. It is also a good substrate for the development of soil microflora. As yet, we can only speculate on the importance of these effects on the soil.

Only about seven species of slug are actually garden pests, for instance the grey field slug (Deroceras reticulatum) with its mottled white/fawn coloration and milky white mucus, is rightly public enemy number one. Other species are predatory, eating other slugs, earthworms, centipedes and even aphids.

Your worm bin will make an ideal home for slugs and snails. Moist, warm, sheltered and full of food, what more could they want? However, unless you can identify and separate out the pest species it is wise to keep them all out of your compost, as you are likely to spread their eggs or young when you come to use the compost. I can identify non-pest slug species and am happy to add them to my worm bins as they speed the process up. I remove all snails, as my identification of them is poor.

Spiders

Again no description is needed here. Spiders have figured prominently over the years in nursery rhymes and folklore, instilling fear with their numerous legs and eyes. They are beneficial creatures and consume garden and household pests at all stages of their lives.

There may be several different spiders in your worm bin. Some will make webs to catch prey, especially around the edges of the bin where there is a solid surface. Others will be hunters, with a plentiful supply of prey. As a rule they are not a problem to the system at all, but if you actively encourage woodlice to speed the decomposition process then you would probably rather the plump, fierce-looking woodlouse spider kept away. With its huge jaws, rusty-red fused head and thorax and yellow abdomen, it spears its victim and injects it with a powerful poison. It does not make a web, preferring to hunt its prey, usually at night.

Woodlice

Around 37 species of woodlice live wild in the UK, five of which are common. They are a familiar sight in many gardens. Closely related to shrimps, lobsters and crabs, they are the only crustaceans to have successfully adapted to living an exclusively terrestrial life. However, they are vulnerable to drying out, so have to live in a damp environment. As they feed on dead and rotting vegetation, wood, fungi, algae and even carrion, your worm bin can provide an ideal environment for them.

2mm

Overall, woodlice are of great benefit to the composting process as they chew up the foods and make it more suitable for microbial activity and worms when they expel it at the other end. I collect woodlice in my worm bins for this very reason, and try to have several different species in each bin, because the different species prefer different foods. I thoroughly recommend putting them into worm units. If your worm bin has a window it is fascinating to watch the woodlice. Occasionally woodlice will eat seedlings, especially when there is no other food source available to them (as in a clean greenhouse), and are thought of as a pest by some gardeners. This would not be a problem in a worm bin.

An interesting snippet about woodlice is that they eat their own faeces, because they need the copper that they contain. The bacterial activity in faeces changes the copper into a form that they can use so they have to eat it again to make use of it. Because of this I add an old rotting log to the bin and the woodlice soon adopt this as home, making it a lot easier for them to find their faeces in an environment continuously turned over by the tunnelling worms.

Other residents and visitors

Your compost heap or worm bin may also be visited by other organisms, especially if outside and on the ground. They may drop in for a feed, to find shelter or to lay eggs. These visitors could include predatory ground beetles, rove beetles, earwigs, harvestmen (daddy long legs), moles, badgers, foxes, birds, rodents, slow worms, grass snakes, bumble bees, wasps, flies and next door's kids! Whether you welcome them or not is up to you - there are hints for keeping the larger ones out in *Chapter 7, Frequently Asked Questions.*

How to make vermicompost 4

"The technique is easy, and involves much less work than ordinary compost-making, and in all seriousness I suggest that everyone turns their attention to increasing the earthworm population (and there is no one who cannot do this, for it can be done even in a flower-pot or window-box)"

Lady Eve Balfour – founder of the Soil Association

When you set up a wormery you are aiming to provide worms with the best environment for them to 'do their stuff'. Make them happy and you will be able to harness their amazing capabilities.

But what many people fail to realise when they first start with wormeries is that they are about to become farmers, and their worms are their livestock. In order to keep their livestock happy, farmers need to understand their requirements - as described in Chapter 2. As well as providing for the needs of the worms, the ideal worm bin should be easy to manage and easy to harvest. If it is difficult for you to manage your bin you are less likely to make a good job of it. Once you have got to grips with the system that you choose you will find being a worm farmer a pleasure. Put simply, give them what they want and they will give you what you want! This chapter will explain how to do just that.

Giving them what they want

To get the best out of the worms, they need a constantly warm and moist, but not waterlogged, environment. They also need ventilation, good food and a partner. And they like it dark.

Providing a suitable home

There are several factors to consider when choosing or designing a worm bin. How much space you have, how much waste you wish to vermicompost, ease of management, etc. This chapter will enable you to make an informed decision about all of these matters.

There are several ready-made bins available commercially, which will be reviewed in the next chapter, and the information in this chapter on setting up your bin will enable you to make an informed choice should you choose to buy one.

The scope is almost endless for building a wormery yourself, either from scratch or adapting a container you already have. Instructions for simple bins are included in this chapter.

Size

You must decide before you either buy a wormery or start to make one, what size you need. This will depend on how much you want to put in it as well as the area you have to site it, including space for working around it.

For the worms, it is the surface area that matters – the larger the surface area, the more they can eat. There is a general rule for how big a bin to go for. Mary Appelhof, pioneer in home vermicomposting, in her excellent book, 'Worms Eat My Garbage', recommends one square foot of surface area for each pound of waste per week ($1/10^{th}$ square metre surface for each ½ kg). If you only want to recycle your food waste you may like to complete a small audit of how much you throw away weekly. Will you vermicompost only kitchen waste, or do you want to include garden waste and cardboard? The amount you vermicompost will vary according to your daily habits and lifestyle and will be individual to you. A bin size based on this information should cater for your household waste.

The surface area in the bin is all-important, but it does not want to be very deep. Anything over 76 cm (2 foot 6 inches) is really wasted and can make life difficult, especially when harvesting the vermicompost. This depth allows the worms 15-23 cm of an active feeding area and the remaining depth used for bedding material. Composting worms migrate upwards to feed in the top 15-23 cm of the material, using the bedding material underneath to escape into, lay their cocoons or just have a break! The deeper the unit, the more likelihood there is of the bedding packing down and compressing, driving out the air and encouraging anaerobic conditions, which will kill worms.

The fabric of the bin

I have seen worm bins made of wood, resin-bonded cardboard, metal, plastic, corrugated iron, and larger outdoor worm beds made of breeze blocks, bricks, wood, corrugated iron and plastic sheeting. Some materials are better than others, and all have advantages and disadvantages. I have constructed wooden bins, adapted plastic containers and polystyrene trays and tested commercial worm bins. Wooden bins win hands down with me every time.

However, because they are cheap and easy to manufacture most vermicomposting bins are plastic. Many commercial bins are simply dustbins or wheelie bins with modifications. These models are cheap and easy to make, so there is a place for them and if they are carefully managed they can provide a suitable home for your worms. Some do have an environmental bonus in that they are made from recycled plastic.

Plastic bins do have disadvantages, though. I have known enthusiastic people put off vermicomposting after buying a fairly cheap plastic bin - a great shame. The bins do not breathe and the air holes can tend to clog up, so the inside can be continuously wet and sweaty. This excess liquid fills the air holes in the bedding leading to anaerobic conditions, which will kill the worms. In addition plastic bins are often dark in colour and absorb and retain the heat from the sun in summer. This can overheat the worms. In the winter the bins are not thick enough to insulate the worms from the cold. Some commercial plastic bins do not have sufficient room for bedding. Mobility of the unit and harvesting the vermicompost can be difficult with some plastic wormeries. However, with careful management - as described later in this chapter - they can make a suitable home for your worms.

Home-made bins

Home-made bins can be cheap, simple and reasonably satisfactory. My first home-made bin was made from a discarded polystyrene fish container, followed by a wooden box made from plywood I had lying around. Later I adapted a plastic dustbin and a plastic waste paper collection box.

They all had faults in their design and in all cases the worms died and had to be replenished - I was on a steep learning curve! The one that did work (albeit with some management problems) was the converted dustbin.

Plastic

Plastic containers can be easily obtained and adapted for worm bins and can give fairly reasonable results, once a few basics regarding wormery management have been grasped.

Modifying a plastic bin to make a wormery

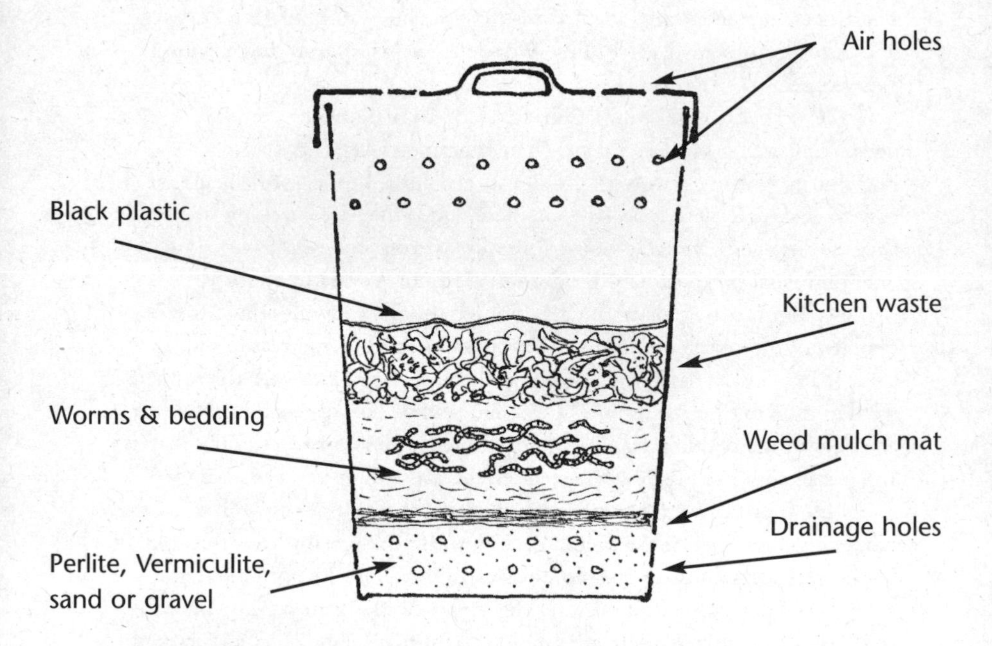

Using an ordinary plastic dustbin and lid, start by drilling drainage and air holes. Make 5 mm holes all over the bottom, about 4 cm apart, and two rings of holes around the side about 7.5 and 15 cm up from the bottom. Also drill 2 mm holes in the lid of the bin and in the top 30 cm of the sides of the bin, again about 4 cm apart (smaller holes to lessen the likelihood of large flies entering). You may find that the holes in the lid let in too much rain during the winter, in which case cover the lid with plastic in the winter months.

Wherever you decide to position the bin, cover the floor underneath with cardboard or several layers of newspaper. This will soak up any moisture that seeps out of the drainage holes and can be periodically replaced, adding the discarded paper to the wormery.

Fill the bin with perlite, vermiculite or polystyrene chips to a depth of about 20 cm. Sand or gravel will do, but they make the bin very heavy, and sand holds a lot of moisture. This layer allows the bin to drain. Cover this layer with horticultural weed/mulch matting, cut to size, to stop it getting mixed up with the worms and the vermicompost.

Your bin is now ready for the bedding material and worms - described later.

To get the vermicompost out of this type of bin you will need to empty it out and use one of the dump 'n' sort harvesting methods (full details later). Remember to take the opportunity to unblock the drainage holes using a metal skewer or similar when the bin is empty.

Wood

Wooden wormeries have a number of advantages. Wood tends to be able to breathe and is a better insulator than plastic in extreme weather conditions, helping to keep the inside cooler in the summer and keeping the cold out in winter - the thicker the wood, the better the insulation. A bin made of wood can be aesthetically pleasing, and wood is easy to work with and allows you the flexibility to enlarge the bin or add a Plexiglas window or door.

On the down side, the worm bin will be continually damp and wood may deteriorate if not treated. Wood can also dry out and therefore will need more water management.

Making a wooden wormery

Basically you need to construct a four-sided, bottomless wooden box, with a roof. Choose the size to suit your own circumstances but bear in mind that broad and shallow is better than narrow and deep.

The box and roof can be made from plywood or planks, at least 20 mm thick. The bottom needs to be strong mesh, and although strong chicken wire stapled to the bottom of the box may suffice, 2.5 cm galvanised mesh would be far superior. The extra expense for the mesh

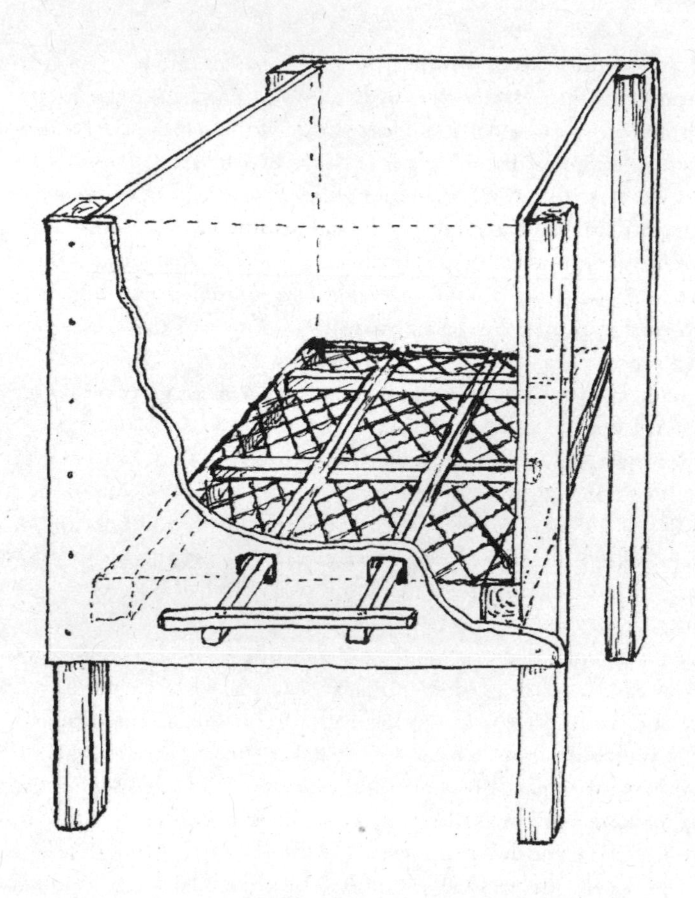

will soon repay itself in ease of use and durability. Include a rim of 2.5 cm batons inside the bottom of the box to support the mesh.

The addition of legs to raise the box off the ground allows air circulation and easy harvesting by the continuous flow method. The continuous flow method is easy to operate and has many advantages. To harvest your vermicompost you could simply (but messily!) tip everything out and dump'n'sort, or take the time to add a harvesting bar. This will literally scrape the vermicompost off the mesh and into a wheelbarrow or tray situated below. Consider this before you decide the height of your legs. They should be made from robust pieces of timber or angle iron, drilled so they can be bolted to the sidewalls. If your wormery is to be indoors you might find adding castors useful.

To make the harvesting bar, cut two batons at least 10 cm longer than your box, and three just shorter than the width. The two longer pieces will run from front to rear and the shorter pieces will be affixed across these at right angles. Fix two of the shorter pieces to the longer pieces to make a square, with sides somewhat less than half the length of the longer batons (see diagram). Now make holes in the front of the wormery so that the long batons will fit through. Fit the whole harvesting bar inside the wormery by inserting the two long batons through the previously made holes in the front. Fix the third short baton to the protruding ends, (which should be on the outside now!) to make a handle. If you are fitting a harvesting bar make sure that you have the mesh the right way up for the bars to sweep smoothly over it, that is the 'front to rear' wires are on the top of the 'left to right' wires.

For the harvesting bar to work effectively, the vermicompost has to be in contact with the wire mesh floor. Occasionally, especially if the vermicompost dries out underneath, it may stick together above the mesh grate. Keeping the conditions inside permanently moist should eliminate this problem.

One of the advantages of building a wooden wormery is that it is easy to add a Plexiglas window, which allows you to monitor the composting process. It is also extremely interesting for children (and adults) to have a view through a window into the 'world of worms'. Do remember to include a door in the design or block out the light by covering with a black plastic sheet or hessian sack, as the worms will not work any material where there is light.

Treating wood

Bear in mind that the worms will be in close contact with the bin, so it must be treated with a non-toxic product. I would not recommend buying timber that has been treated with copper, chromium or arsenic (CCA) purely for environmental reasons, nor creosote or a similar product for the sake of the worms. Certain paints are made of oil and other chemicals that could prove toxic to the worms over a period of time. Treat the wood with an environmentally friendly preservative waterproofing compound or other non-toxic coating. Raw linseed oil is fine for treating wood, but does lose its water-repellent properties after 3-6 months and would need to be reapplied. Emptying the bin occasionally and allowing it to dry completely then recoating the inside would extend its life.

I do have a wooden wormery that lives outdoors all of the year in an unsheltered location. I treated it just once with linseed oil some four years ago and it has never been treated inside or outside since. Although looking a little ragged along the outer edges, it is still going strong. Whatever you use, leave it for a few days to dry completely - oils on a worm's skin may interfere with its breathing.

Outdoors worm beds or pits

If you feel that you need a large bin and have the space, you can build an outdoor worm bed. They are basically long narrow open-top boxes, similar to raised beds in an organic garden. For ease of use they should be no more than one metre wide, allowing you to work comfortably from either side of the bed. They can be made of breezeblocks or timber, or whatever other strong and suitable materials you have to hand, with a height of two or three breeze blocks. The length of the bed depends on how much material you wish to vermicompost and how much time you can put into it. For a long bed I would suggest you subdivide into smaller units, say 2 m long, with wooden planks for dividers. Of course, if you have a suitable structure in your garden already you can convert it - I have seen old pig sties, made from breeze blocks, put to very good use as large wormeries.

In the colder parts of the United States the worm pits are dug into the soil. The colder the climate, the deeper they dig. Thus in really cold areas a pit would be dug two breeze blocks deep, lined with a wall three breeze blocks high. Being below ground level the pit is easy to insulate in the winter and the contents stay cool in the summer.

Flooring

The bottom of the bed should be covered with gravel and coarse sand to a depth of about 10 cm, to provide drainage. Wooden planks placed on top of the gravel with small gaps between prevent the likes of moles finding a ready food source, and saves you digging into the gravel when you harvest the vermicompost. If moles are likely to be a major problem, chicken wire fixed on top of the planks should really keep them out.

If your bed has a concrete floor it will absorb water, as well as any smell this water may have. Any leachate produced will be absorbed. Once saturated the concrete may well wick water from the ground. The ground water may be slightly acidic causing a chemical reaction, leaving a crystal-like deposit. This will eventually weaken the concrete, assisted by water freezing and expanding in hairline cracks in the concrete. Sealing the concrete will add longevity to the floor. However, a concrete floor will certainly keep moles out!

Roof

Just like a smaller wormery, excessive rainwater and sunlight should be kept out of your worm bed, so make a roof from whatever is cheapest and most effective. Wooden planks would be OK, but a sheet of plywood would be better as it has no gaps for rain to pour through. Corrugated iron would tend to heat up in the summer months. The roof should be kept in position permanently except for watering, feeding and harvesting. Birds, foxes and even badgers would appreciate you leaving the roof off. For the winter months the roof should be insulated with a layer of straw, tied down tightly with a tarpaulin cover - you don't want straw all over the garden.

Siting the worm bin

Now, you have a fantastic worm bin ready to go, where should you put it? Worms need moisture to live and thrive, and reasonably constant temperatures. From a worm's point of view, a good site for your wormery will not be too hot or too cold, or too variable. Researchers have found that our main composting worm, Eisenia fetida, cannot survive exposure to 0°C for 48 hours in a laboratory; so steps should be taken to ensure that your worm bin does not freeze. It would be fair to say that the optimum temperature for composting worms would be between 15 and 25°C, so we should aim for this sort of range inside the worm bin.

The siting of the bin will affect your life as well as the worms, and as I said earlier, if it is easy for you to look after your worms then you are more likely to make a good job of it. Convenient, easy access to the wormery to recycle your food waste is important as you do not want to get your feet wet when feeding your worms.

Wherever you decide to put your bin, you will need access to water and space to work. During very hot weather, particularly if your bin is in direct sunlight, your worms may need several cans of water a day, which you would probably rather not carry too far. Also you will need to empty the wormery, maybe move it around, and harvest the vermicompost, all of which will be far easier with adequate space around the bin.

Indoors

Should you put your worm bin indoors or outdoors? Worms prefer constant temperatures, so an area that is least affected by changes in temperature would be best suited for them (and you). Indoors has several advantages. The temperature is less likely to fluctuate suddenly and it can even be controlled. Inside is fine but over the summer it could get too hot and require more water management. Indoors in the winter, central heating can leave us with a false sense of security and many people forget how quickly the wormery can dry out in these conditions. If the wormery is always kept moist, the worms will keep recycling your waste efficiently all year round. Being indoors, a wormery can provide an interesting topic of conversation. I know of one lady who keeps hers in the kitchen. Flies could become a nuisance with a wormery sited indoors, but they are quite harmless in themselves.

Almost outdoors

I prefer to keep my bins in our brick-built garage. It is cooler in the summer and warmer in the winter than outdoors. One winter I measured the outdoor temperature as -11°C, but inside the wormery, in the garage, it was 4°C, a big difference. Basements and cellars would be good places to keep a worm bin, as would a shed or any kind of outhouse. Greenhouses are likely to become too hot in the summer, and the wormeries would need insulating during the winter if no heat is provided.

Really outdoors!

Outdoors, wormeries need to be kept in a sheltered position. That is, offering shelter from the summer sun, from heavy rains and cooling winds, not to mention frosts. And spare a thought for yourself; choose a site with easy access. Are you really going to walk down to the bottom of the garden in your slippers and dressing gown when the box of kitchen waste is full and needs to be fed to the worms, whatever the weather?

Insulation of wormeries

Wherever you choose to site your bin, you can help to keep the temperature inside reasonably stable by insulating it. This will be particularly beneficial in winter, when the worms slow right down but you are still generating food waste.

Lots of things could be used to insulate your wormery. Carpets, cardboard boxes, old pillows and duvets, even plastic bubble wrap tied round the wormery would help. As an added bonus many of these things could be composted themselves when the weather improved. Bales of straw placed around the wormery would provide excellent insulation. If you want to be more sophisticated you could use pallets to build a box around your wormery and stuff it with straw, shredded paper or bags of polystyrene chips. Polystyrene tiles could make a great outer box too.

Food as an insulator

Some people pile food onto the top layer of their wormeries to act as an insulator and food store for the worms over the winter. The food, because of the external temperature in winter, will decompose extremely slowly, as it is too cold for microbial activity.

I think it would be a better idea to mix thc kitchen waste with a high carbon source, such as wood chips, which may generate a little heat. This is more likely to work indoors than outside.

If you want to use food as an insulator, bear in mind there will come a point when you need to clear up the mess!

Bedding

Worms and worm food go in the bin, but just as important, and easily neglected, is bedding for the worms. We don't want to live in our food, and why should they?

Starting a wormery without bedding is not recommended - the initial bedding is key to success. The waste food could contain too much water and nitrogen for worms and possibly salt. Depending on the food and the amount, it could even generate enough heat to kill the worms. The bedding acts as a buffer zone for the worms to escape extremes of temperature or other unsuitable conditions.

Bedding materials

A number of things can be used for bedding, but some are better than others. The bedding must:

- allow airflow and retain moisture
- be a neutral pH
- be non-toxic to the worms and not overheat

Having used many different bedding materials, I have no doubt that freshly worked vermicompost is the best thing to use.

Advantages of vermicompost as bedding

It's obvious really, isn't it? Worms will be happy to live in vermicompost. It has already been worked by worms, they were born and bred in it, it is already inoculated with the millions of the microbes needed to start the decomposition process and the worms can start feeding immediately instead of having to process other materials to make them into suitable bedding. Worms travel well when packed in vermicompost. Introduced to their new living accommodation, complete with a layer of familiar (and comforting) vermicompost, the transition should be stress-free for the worms. Using freshly worked vermicompost as the initial bedding material is almost as important as the initial worm numbers.

Reputable suppliers can supply enough bedding material to start your worm bin off when you purchase worms or a wormery.

Other bedding materials

Other materials that can be used for bedding are damp shredded newspaper, damp shredded computer/office paper, damp cardboard (especially corrugated cardboard), leaf mould, chopped up straw, de-salted seaweed, wood chips mixed with any of the above, hemp bricks, older vermicompost, old garden compost, aged manures and composted manures. Composts must be past the hot stage of decomposition and under no circumstances incorporate materials that include herbicides or pesticides. A mixture of materials is preferable to any one used alone as they provide more nutrients for the worms and create a richer vermicompost. Some people recommend peat and coir, but for environmental reasons I do not.

Condition of bedding

The bedding should be soaked before use and excess water allowed to drain away before placing it in the worm bin. A general rule of thumb is if you can wring water out it is too wet, if you can't it is too dry. If just a couple of drops come out when squeezed then that is acceptable. It must be loose and friable so as not to restrict airflow and placed so as not to pack down. Adding wood chips will aid aeration. A couple of handfuls of sand or soil in the bedding will provide grit necessary for the worm's digestion of food.

Depth of bedding

How much bedding you need will depend on the method that you use to feed your worms, but I would suggest a minimum depth of 10-15 cm if top feeding. For pocket feeding (see below), the food needs to be surrounded by bedding so you will need a deeper layer, filling the bin about three-quarters full.

Continuous flow usage

If your bin is designed for the continuous flow method, it has a mesh floor. It might be obvious, but put a layer of newspaper or cardboard in the bottom before the bedding, to stop it dropping straight out. It will break down before you are due to begin harvesting

Worms

How many worms?

The more worms you have when you set the wormery up, the quicker the system will settle down and produce vermicompost and deal with your food waste.

Your unit may come supplied with worms, but I am a firm believer in starting with as many as possible. In 2-3 months, given the right environmental conditions worms can double their numbers. So, if you are supplied with 150 composting worms (which is what some suppliers provide), given the right conditions in 3 months you could have 300! Supplied with 2000, you can see what a difference it will make after that short period. This could have an effect on the success or failure of the worm bin, and your patience in wanting to see some results.

Having said that, it is the worm biomass, i.e. the actual weight of worms, that counts, not the number of worms. One pound of worms will eat the same amount of food no matter how large or how many there are. As a rough guide 1 kg of worms will deal with 0.5 kg food. So a good way to work out how many worms you need to start your bin is to weigh your waste for a week and buy worms that weigh twice as much. As the wormery matures, all being well the worms will breed and process more material. You should only need to buy the worms once, so buy as many as you can afford to kick-start the system.

Where to get your worms?

There are many worm breeders in Britain, many of whom advertise in the pages of angling magazines. There are reputable worm suppliers, but believe me, there are many worm breeders out there who do not know one end of a worm from another, nor which species is best suited for a given purpose.

A good supplier will know about the worms and can be very helpful. The health of the worms, age, how they are handled, how they are harvested, speed of delivery once sent out, the packaging, air holes, bedding materials and moisture content all have a bearing on the success of your new arrivals. Your parcel may have to spend time in a delivery depot and the packaging can be critical. Personally I think it would be a wise investment to pay for a well-packaged parcel of healthy worms.

Specifying and receiving the right species is another important consideration. Specify the scientific name(s) of the worms you want and the reason you want them. Do not let the vendor convince you that his species is better for you.

■ *See Appendix for stockists of worms.*

Introducing the worms to the bin

Once the bedding is in the worm bin, gently put the worms on top (of the bedding material). Leave the wormery lid open and completely uncovered to the light, and the worms will move down. You may wish to lightly sprinkle water on top of the worms. Leave them to settle in for a few days. Leaving the lid off can help here.

After this settling in period, start feeding the worms (see facing page) and cover the top of the bin with a black weed mulch mat or a loose piece of black plastic (this will help to retain moisture in the top layer). Close the lid, if supplied.

Feeding

Worms will eat all sorts of things, such a wide range that it would almost be easier to list what they will not eat. An A-Z guide is in the Appendix (worm foods).

As well as things the worms eat, it will help to keep the bin healthy if you add carbon materials to every feed. Green plant materials of course contain carbon, but it takes the microbes 30 carbon molecules to use one nitrogen molecule. We can provide the right balance by adding carbon-rich

materials such as straw, cardboard, wood chips/shreddings and paper.

There are several methods for feeding worms in a worm bin. The method you use will depend on the style and size of the unit and how you wish to manage it.

When and how to feed

For all systems the feeding rate will depend on how quickly the worms consume the material, time of the year, number of worms, etc. But the rule of thumb (which we used earlier to choose how big a bin you wanted) is ½ kg food per $1/10^{th}$ square metre surface area per week.

Calculations aside, the worms will let you know when they want more food. How? When you lift the lid, if you see just a few worms in the top of the food/bedding it is the time to add more food. No worms on the surface, no more food. Simple! This is the case with all of the feeding methods.

Whatever feeding system you use, a good tip is to cover your food waste with a 2.5 cm layer of fresh vermicompost. This has several advantages. Your food waste is 'sandwiched' between a top and bottom layer of vermicompost which means that decomposing microbes can start from the top and the bottom. It really does speed the whole process up. Secondly, it can help to keep flies out of the food waste, particularly fruit flies that just want to lay their eggs quickly and not waste energy and time trying to find a way through this top layer to the rotting food below. The top layer also helps to keep the food moist and prevent it from drying out. If you don't have any vermicompost yet use damp shredded newspaper, damp shredded cardboard, damp leaves, leaf mould or even wood chips.

Top feeding system

This is the feeding method that I choose to use in my home wormeries. If the design of your bin allows you can use the 'continuous flow' method, which combines top feeding with an easy harvesting method (see below).

The worm bin is started with a 10-20 cm layer of damp bedding material (preferably fresh vermicompost). A 2.5 cm layer of food is placed on top of the bedding, on top of which I then put a top layer of vermicompost as described earlier. As the worms eat the food another layer is added, and covered by a damp top layer. Slowly the layers build up. When the bin is almost full, the worms will be working in the top 15-23 cm of the bin and the material underneath can be harvested. If you are harvesting with a dump 'n' sort method, the top layer should be saved for 'seeding' the next cycle.

Pocket feeding

Using this method the bin is partially filled with damp bedding and the food is completely buried in the bedding, in pockets. Different areas of the worm bin should be used for pockets - the bigger the worm bin, the more pockets. Food is added regularly, rotating the pocket feeding sites. Harvesting occurs when the unit is full of vermicompost and no recognisable bedding is visible - unless you started with vermicompost as the bedding, in which case everything will look the same. The whole of the unit is emptied, worms and all and must be sorted.

Pocket feeding is a useful method for large wormeries where there is space for plenty of pockets, although it is possible to use it in smaller wormeries too.

Stacking Trays - or Upward Migrating Worm Composting System

There are several commercially available wormeries with trade names such as 'Can'o'Worms' or 'Worm Tower' which are based on a system of loosely fitting trays stacked one above the other. The trays have holes in the bottom, like garden sieves and form a tower of separate trays.

There is usually a liquid collection area in the bottom tray, and legs raise it off the ground. These bins are usually made of black plastic, which can cause overheating problems in the summer. The basis of this system is that worms will move upwards into fresh food. The vermicompost is formed in a series of layers in the bottom tray and as the worms finish one tray they move up into the tray above it. This was designed to minimise hand sorting of worms from the vermicompost. Harvesting the vermicompost means removing all of the trays to reach the contents of the lowest tray. These are neat systems that can work well when managed carefully.

To start the wormery off, bedding (usually supplied as a coir or hemp brick) and worms are put in the bottom tray, and food is gradually added until that tray is full. Food is then added to the empty tray above. The idea is that when the worms have finished working the food in the lower tray they move up into the food above, and the lower tray can be removed, emptied and used again.

In practice I have found that the worms tend to stay in the lower tray, which is more microbially active and therefore rich in food for the worms. When the lower tray is removed there are usually lots of worms to pick out by hand. Even once the lower tray has been completely worked by the worms, they often cannot get into the tray above because the volume of material or food deceases as it rots down, the worms eat it or it simply settles. Wherever this material touches the upper tray it acts as a bridge between the bottom and upper tray. However as it decreases a gap is formed between the trays, making it very difficult for worms to move from the lower tray to the upper tray. Worms cannot jump upwards - yet! You can overcome this problem by pushing some material higher or crumpling some damp newspaper so that it forms a bridge.

Being plastic, these bins sweat a lot and the liquids accumulate in the bottom container, where worms can drown. Draping a large piece of permeable material, such as a weed mat, over the bottom chamber and underneath the lower tray will keep some of the worms out of the reservoir below.

The liquid in the sump can start to smell but leaving the tap open with a bucket beneath to catch liquids will help. Unfortunately this may allow flies in so tape some weed mulch mat to the tap to keep them out but still allow the unit to drain.

If the airflow in the bin is not adequate, anaerobic conditions can easily take over. If the unit has air holes it is important to regularly check that they are clear. If it doesn't you may consider adding some yourself.

A good tip is to add bedding to each new tray before you start adding your food waste, otherwise once the lowest tray is removed the worms have no bedding material at all in the unit and have to live in their food.

Lateral movement system

Usually associated with large outdoor worm beds rather than bins, the lateral movement system basically involves splitting one large bed into two separate working beds. It is a simple method for the tricky and somewhat time-consuming problem of separating worms from the vermicompost.

The bed is divided into two with a large piece of wood or similar. Bedding, food and worms are added to one side, leaving the other side completely empty. The worms work their side and when they are nearly finished, bedding and food are added to the second side and the divider removed. Having eaten all the food in the first side, the worms migrate to the new food on the other side and work that. This could take a number of weeks, but with the end result that the vast majority of the worms would have migrated across to the 'new' bed. The divider would then be replaced and the vermicompost in the first side harvested. The process is reversed ad infinitum.

This method does allow flexibility. Depending on the actual depth of the beds, you could either top feed or pocket feed continually until the bed is full, or then let the worms roam next door. If your beds are not deep enough to allow pocket feeding, the top layer feeding method would be best, continually adding food in layers.

A simple variation on this method is to start with the whole bed but when the majority of the food has been worked, scrape the vermicompost to one side and prepare the other side with fresh bedding and food. The worms will move into the new side. The newly prepared side now becomes the working side with additional food added to that side only.

There are several advantages to the lateral movement system. The vermicompost is left in the bed for an extended period of time, allowing the worms to process it thoroughly. There is no manhandling of the worms during harvesting of the vermicompost, so they are not hurt or put under any unnecessary stress. They are free to move from one side to the other if conditions on one side become inhospitable. Also it saves a lot of time, not having to physically separate the worms from the vermicompost. You do however require a larger surface area, because in effect you have two (or more) working bins, side by side. There could also be a tendency for the non-working side, if food is added too soon, to become anaerobic and smelly whilst the other side is still being worked.

Creature comforts

In the wild, worms may find temporary shelter from temperature extremes, humidity and predators by using logs, stones, and crevices in the soil, burrowing deeper into the litter or hiding under the bark of trees. Such areas are not available to them in a worm bin, but I feel it would be to their liking if there were such places for them to just chill out. I place a large piece of timber, about 25 cm square and 20 mm thick, on top of contents of my worm bins, and after a very short period the area underneath the timber is teeming with worms (and the woodlice I put in my wormeries also appreciate it) Perhaps this is mimicking their natural behaviour?

Harvesting the vermicompost

We've fed the worms and fed them some more and finally the bin is nearly full, so now it is time for us to harvest the vermicompost.

Harvesting basically means separating the worms from the finished vermicompost. The method you use will depend on why or what you wish to harvest. I tend to think of the vermicompost as the finished product and save the worms for the next batch in the bin, but if you are a worm breeder or fisherman the worms will be your primary harvest.

For those of us that are harvesting the vermicompost, it will be none the worse for having worms and cocoons in it. You do want as many worms as possible in your 'seeder' for the next new batch, though, so the more worms and cocoons you remove from your harvest the better.

Some wormeries (such as the wooden design at the beginning of this chapter) have been designed for the 'continuous flow' method of management with an easy, no fuss harvesting method. Unfortunately, if your bin does not have this feature the worms and vermicompost have to be completely emptied out of the wormery for harvesting. In this case, use whichever dump 'n' sort method you prefer. Apart from the continuous flow method, all of the harvesting methods are basically refinements on dump 'n' sort. You will need a container in which to put the worms and cocoons that you collect.

Dump 'n' Sort

Literally, tip the contents of the bin onto the floor and, one way or another, separate the worms from the vermicompost. Time consuming and messy! Not such fun for the worms either, who literally have their world turned upside down.

Some units are easier to tip out and empty than others, with large units being particularly difficult simply due to their size. Remember that if your bin has a sump it may well be full of liquid This will have to be emptied first as you don't want the liquid soaking your vermicompost or spilling out onto your garage floor. Scooping the contents out of the wormery would be much tidier than tipping it out.

We can use what we know to ease the sorting process; we know that the majority of the worms will be found in the top 15-23 cm of the worm bin. If we separate that top layer into a large container, ready to use as the bedding for the next batch, there will be fewer worms in the remaining vermicompost, making hand-sorting the worms and cocoons much easier.

Dump 'n' Sieve

This variation on dump 'n' sort uses a sieve to catch many worms and cocoons missed when hand sorting the lower layer of vermicompost. Again, separate the top 15-23 cm of vermicompost, where most of the worms are, to start your next batch. The remaining vermicompost is hand sorted to remove worms and cocoons, then sieved though a 5 mm garden sieve into a large plastic tray (the lid of a dustbin works well). The worms that are left in the sieve are added to the starter bedding for the next batch.

Attract 'n' Trap

This method requires that you do not feed the worms for a few days, so that they will move into fresh food. When the worms are hungry place a large garden sieve, or an onion bag or something similar on top of the worms and bedding. Put fresh food into the sieve, cover with a thin top layer of damp leaf mould or damp newspaper and leave for a few days. The hungry worms should move into the new food. You can now remove the sieve, which should contain most of the worms. The remaining vermicompost needs to be dumped 'n' sorted or sieved as described above to remove any remaining worms and cocoons. This method has been especially successful when fruit, particularly watermelon rind, is used as the food to attract the worms.

Cone/Pyramid

This is another modification of the dump 'n' sort method, and a very successful one, which I used before I had continuous flow bins. It works best either outdoors in direct sunlight or indoors with the lights on.

Remove the top 15-23 cm of vermicompost, where most of the worms are, and use for your next new batch. Cover an area of ground with a large sheet of plastic, and put damp newspaper on the middle. The entire contents of the wormery are tipped on top of the damp newspaper. The wormery contents are piled into a cone or pyramid shape. Worms, not liking light move into the pyramid. Leave the worms for a while then gently scoop away a few centimetres of vermicompost from the top and sides. Reshape the mound into the cone/pyramid and repeat the procedure; wait, scoop, shape. Eventually as more and more of the vermicompost is harvested the worms will take refuge in the damp newspaper, and this can be used to 'seed' the next new batch.

Remember to separate the worm cocoons from your harvested vermicompost - they cannot move away from the light for you. You can pick them out of the vermicompost while you wait for the worms to move deeper into the mound.

A quicker variation of this method is to use a larger sheet of plastic and make several cones/pyramids instead of one.

Continuous Flow System

This is more than a harvesting method; it is a cycle of feeding and harvesting, with small amounts of food going into the top of the bin and vermicompost coming out at the same rate at the bottom. It eliminates the need for separating the worms from the vermicompost, by harvesting it from underneath the active layers. The top feeding method is used, or even pocket feeding can be used with the larger units and, unlike the other harvesting methods, the bedding should never need to be replaced.

These systems have raised floors that can be made of metal grills or even very taut ropes. Harvesting involves disturbing the vermicompost at the bottom with a rake or by using the built-in harvesting handle, so that it falls into a tray or onto the floor. Only about 2.5 cm (1 inch) of vermicompost should be harvested at a time, keeping a large volume of bedding in situ.

The continuous flow system does require some time to set up properly. Harvesting should not begin until the bin is nearly full of vermicompost.

Feeding in regular, thin layers will encourage the worms to concentrate in the upper areas of the system, where microbial activity is highest. Over or underfeeding will tend to encourage the worms to migrate throughout the material in the unit in search of food. This might result in worms being harvested with the vermicompost, and will lessen the efficiency of the process.

Continuous flow systems are the easiest currently available to manage and harvest. Having spent years using different systems, I would always recommend them. The key factors are:

- to keep adding thin layers of food at the top
- to build up a deep volume of vermicompost as bedding before harvesting small amounts from below
- keep the whole system moist

Stacking trays

These operate as a sort of 'continuous flow' system in that you add food to the top tray and harvest the contents of the lowest tray. When all the food in the lowest tray has been eaten, it can be lifted off and sorted. The idea is that when the worms have finished working the food in the lower tray they move up into the food above, and the lower tray can be removed, emptied and used again on top.

In practice some of the worms can stay in the lower tray so it is probably worth doing a dump'n'sort to put any lingering worms back into the system. The trays can be heavy so take care when you lift them off.

Keeping it all going

In an ideal world, you would feed your worms and when the bin was full you would begin to harvest regularly, with everything going along smoothly. Life is not always that simple, so don't be dispirited if it doesn't work perfectly first time. Worm composting at home is an art not an exact science. As long as you understand the principles of vermicomposting you will soon be able to manage any problems that arise.

Moisture in the bin

Moisture is of the utmost importance to worms; without it they will shrivel and die. Equally, with too much liquid in the bin they will suffocate. This careful balancing act has to be right, for the sake of the worms.

If you want to check your bedding, use the 'squeeze test'. If you can squeeze more than a couple of drops of liquid from the vermicompost or if it stays in a balled clump instead of breaking apart, it is too wet. If you blow on the surface of some the material and dust blows into the air, it is too dry.

Not enough water

If the worms become too dry, they will lose body weight and reduce their rate of feeding, reproduction, etc. and in the worst case, they die. Worms may recover from a period of dryness, absorbing moisture from their surroundings, but they will not be performing at their best if they are subjected to this sort of stress.

To add more water use a watering can with a rose and gently water the bedding and food, allowing it to soak in. Depending upon how dry the bin is it may take several can loads to completely moisten the contents. You may have to water more in summer than in the winter. Remember indoor bins will dry out quickly when the central heating is on. Adding materials like newspaper, woodchips and cardboard will help to retain moisture. Be sure not to overdo the watering - that causes other problems.

Too much water and water logging

Small gaps in the bedding and vermicompost in the bin hold air, enabling the worms and microbes to 'breathe'. If the bin becomes too wet (which it can do just from the moisture contained in the vegetable waste you are adding), water fills the air gaps, driving out oxygen and leading to anaerobic conditions (see below). Signs that the bin is too wet are excess liquids at the bottom of the unit (depending on the system you have) or a nasty smell coming from the bin, and in the worst cases numerous dead worms.

Some vermicompost bins have reservoirs or sumps at the bottom to collect excess liquid. Such liquid should be drained off before it can soak up into the vermicompost. Other bins, with open bases, allow excess liquid to drain away, so over-watering is less likely.

If the bin is too wet, the material needs to be well aerated by opening it up, mixing with dry straw, wood chips or corrugated cardboard, Do not add any more food, especially kitchen waste, until the smell disappears.

What can go wrong?

Run properly, a worm bin smells of musty earth and can be rather pleasant. Many people stick their heads into my wormeries and say 'Oh, it doesn't smell!' A bad smell can be one of the first signs that something is wrong.

I must add here that wherever I say use dry, carbon rich, materials, without doubt wood chips are best. Since I have been using them my vermicomposting has improved greatly and I have suffered fewer problems.

Anaerobic conditions

The most common fault with wormeries (and compost heaps) is allowing them to become anaerobic, which will kill the worms. It is without doubt the most difficult problem you will come across, and the main reason why so many 'would be' vermicomposters give up. However, it can be relatively easy to remedy. Do not despair and chuck the whole lot out - you can soon restore the health and vitality of your worm bin.

Anaerobic conditions arise when the worm bin is too wet, and liquid fills the gaps in the bedding and vermicompost, driving out oxygen. Worms will move away from areas where they cannot breathe, and aerobic microbes (the good ones) cannot survive without oxygen, but anaerobic ones can, and thrive. Fungi, which are also very important decomposers, will die in anaerobic conditions.

The result of anaerobic composting is a messy, smelly compost and dead or dying worms. The by-products can include volatile organic compounds, such as phenols, terpenes, butyric and valeric acids, various alcohols, leachates with wonderful names like cadavarine and putrecine, and gases such as hydrogen sulphide (the bad egg smell) and methane. This 'leachate' mixes with water given off from the decaying vegetable waste, and any water you add to the system. It moves slowly down through the contents of your wormery which is decomposing at different rates. As it trickles down it carries with it microbes, soluble nutrients, particles of decaying organic matter as well as vermicompost.

Leachate

Thus you end up with a rich nutrient soup in the bottom of your wormery that may include a whole range of real nasties. The anaerobic microbes that trickled down with it continue to work and thrive in this airless soup. The resulting smell is very unpleasant and acidic. It can be a toxic soup for worms, beneficial organisms and plants. Alcohol and phenol are troublesome both during and after the vermicomposting process and can remain so for years. Pathogenic microbes and phytotoxic compounds may well be present, and they may not have been neutralised by being consumed and passed through the gut of a worm.

What do you do with this liquid that is rich in plant nutrients but potentially harmful? I have heard it called 'worm tea' by so-called experts, who recommend its use as a plant food. I could not recommend this, and people that do are ill advised. All recent research on leachate advises against its use in this way. There is no way of knowing whether a particular leachate is going to have a positive or negative effect. A domestic wormery or compost heap never produces a consistent product and it may harbour compounds and organisms that can harm your plants and the soil micro-organisms around them.

Leachate, the liquid that drains from an active worm bin is not worm tea. You can make a very good and reliable worm tea from your worm bin when it is healthy but it is a very different process - see Chapter 6 for full instructions.

What should you do with leachate?

Unless your worm bin is saturated and therefore does not need any more moisture, pour the leachate back into the worm bin (otherwise, pour it onto your compost heap). I would add some wood chips or shredded cardboard, especially if there is a lot of liquid. If it seems odd to put back a foul smelling leachate into your precious wormery or compost heap take heart from the fact that it is still full of microbial activity and plant nutrients and you might as well make good use of it. Also anaerobic material such as leachate will be made aerobic with the reintroduction of oxygen. Some of the aerobic organisms in the worm bin or your compost heap can actually deal with the toxic compounds rather quickly, rendering them harmless to the worms and the system.

How to remedy an aerobic worm bin

All is not lost if your worm bin does become anaerobic. It may smell and be very wet but it can be made aerobic with the introduction of oxygen. Many 'experts' recommend the addition of lime, calcified seaweed or a de-acidifer but this is not the answer - what is needed is oxygen. Turning and opening up the contents will help, as will adding carbon in the form of hefty dry materials, like straw, corrugated cardboard, wood chips and shreddings (which take a long time to break down) and, to less effect, paper. These materials soak up the moisture and allow more oxygen into the heap. Once the system is oxygenated, aerobic microbes rapidly break down the toxic compounds left behind by their anaerobic cousins. Thus, the alcohol and phenols do not remain a problem for too long, provided oxygen is allowed to flow through the heap and the aerobic microbes are allowed to do their work.

Remember once the bulky, carbon-rich material has been added, leave the unit alone. Do not feed until the smell has disappeared and worms are actively on the surface of the bedding searching for food. This 'healing' process may take some weeks depending upon how bad the problem was.

An anaerobic wormery is probably the only major problem that is likely to arise. There are other issues such as more flies than you would like to see that are covered in Chapter 7, Frequently Asked Questions.

Review of commercially available worm bins 5

"Man has yet to invent, devise or manufacture any machine, any solid or liquid fertilizer as efficient as the earthworm."
Friend Earthworm – Practical Application of a Lifetime Study of Habits of the Most Important Animal in the World by George Sheffield Oliver

Commercially available bins

A range of ready-made worm bins is available in many sizes and different materials. There are three main systems of management:

- Adapted wheelie bins or rubbish bins
- Stacking bin or tray systems with upwardly migrating worms
- Waste buster or continuous flow management systems

This is an overview of the more popular types of currently available new worm bins are being developed all the time.

Companies may offer varying numbers of worms, sundries, value packs or special offers. I would advise that you telephone or visit their web sites prior to purchasing. The information detailed here was correct at time of publication.

The Waste Buster

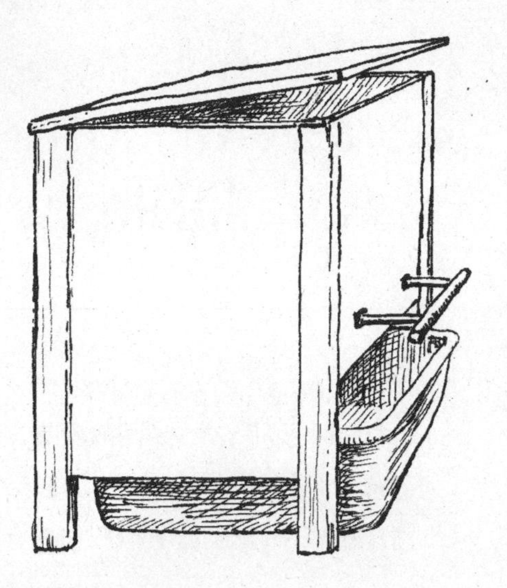

At a glance:

- 'Continuous flow' design
- Made from thick, Forest Stewardship Council (FSC) wood, coated in natural linseed oil
- Various sizes available, from 2 people to small businesses
- Can be supplied with worms and a specially developed bedding
- Harvesting is very simple - just push 'n' pull the handle
- Very easy to manage and use
- Neat and tidy looking system
- Compost is harvested into a tray provided
- Liquids, commonly found in plastic units, do not present a problem
- Good aeration
- Delivered as a flat pack for ease of transport, and easy to assemble
- Heavy duty galvanised metal harvest bar and grill mesh floor
- Each unit stands at least 640 mm high
- Option of a large plexiglas window fitted
- Optional castors can be fitted
- Ideal for educational purposes
- Manufactured in the UK
- Designed by award-winning company
- A comprehensive and thoroughly researched manual comes with the unit

Stockists: **Nurturing Nature Ltd, Worms Direct**, and the **Recycle Works**

Waste Juggler

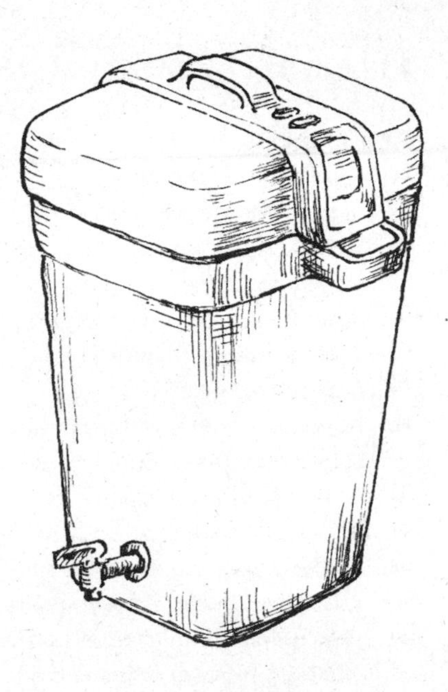

At a glance:

- Large wheelie bins that have been adapted for use as worm bins
- Harvesting is dump 'n' sort
- Large size (90 litres) means that it can hold a large volume
- Height is 800 mm, width and length are 450 x 450 mm
- Easy to move around with sturdy handle and wheels
- Harvesting can be difficult, as the whole unit has to be emptied
- Neat all-in-one design with clip-on lid
- Block of bedding and worms supplied
- Woven bin liner to put inside unit
- Drainage chips and a tap to drain off liquid
- Options may include anti-acid lime mix, moisture mats and worm treat
- Comes complete with instruction booklet

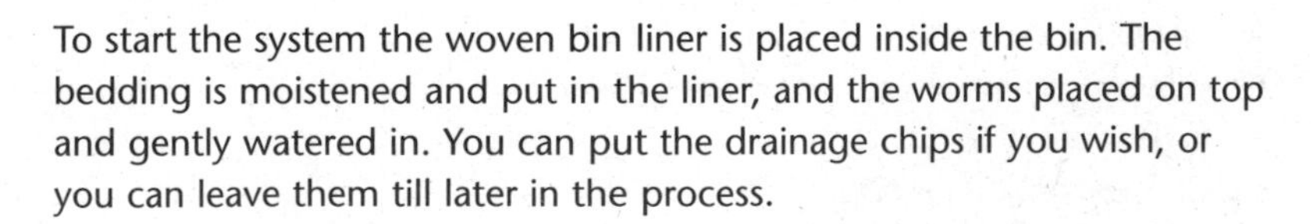

To start the system the woven bin liner is placed inside the bin. The bedding is moistened and put in the liner, and the worms placed on top and gently watered in. You can put the drainage chips if you wish, or you can leave them till later in the process.

Stockists: **Wiggly Wigglers**

Original Organics – Original Wormery & Junior

At a glance:

- ➥ Modified rubbish bin
- ➥ Simple and neat design
- ➥ Harvesting is the 'dump 'n' sort' method
- ➥ Widely available
- ➥ Manufactured from tough polythene
- ➥ 762 mm high, 610 mm wide and 432 mm deep
- ➥ Suitable for a family of four
- ➥ Comes complete with 450 native composting worms, bedding, anti-acid mix, internal drainage platform, a tap and full instructions
- ➥ A double-sealed lockable lid
- ➥ Lid contains small brass air vents
- ➥ Tap to siphon off liquid
- ➥ An insulation jacket for the winter months is available
- ➥ The Junior is a smaller version of the Original Wormery - approx. 440 mm high by 340 mm wide at the top

Stockists: **The Green Gardener**

BritishEco – Wormery

At a glance:

- ➥ Modified plastic waste bin
- ➥ 200 litre capacity
- ➥ 750 mm high x 600 mm in diameter
- ➥ Lid
- ➥ Tap for siphoning liquid
- ➥ No legs - place directly on level ground
- ➥ Bedding, worms and booster mix supplied
- ➥ Harvesting is the dump 'n' sort method

Stockists: **British Eco**

The Green Bin

At a glance:

- Modified rubbish bin
- 50 litres volume
- Place directly on level ground, no legs
- Lockable lid
- Tap for draining liquid
- H500 x L400 x W400 mm
- Worms and starter kit supplied
- Harvesting is the dump 'n' sort method
- 30 litre work top version also available

Stockists: **West Country Worms**

The Worm Hotel

At a glance:

- Stacking Bin System (Upward Migrating Worm Composting Bin)
- Box-shaped tray system
- Made from recycled black plastic
- The standard unit comes with 2 working trays with 46 litre capacity
- Can purchase just one working tray
- Trays can clip together
- Separate liquid collection tray
- Tap for siphoning liquid
- Drainage blanket/mesh
- Snap-on lid
- Has no legs – stands on a level floor
- Supplied with worms, bedding, anti-acid mix, worm nibbles, worm rafts and instruction booklet
- Manufactured in the UK
- Harvesting may need the dump 'n' sort method

Novel approach to prevent worms from drowning in the sump – worm rafts!

Stockists: **Worm Hotel Limited**

Can 'o' Worms/ Deluxe Worm Tower

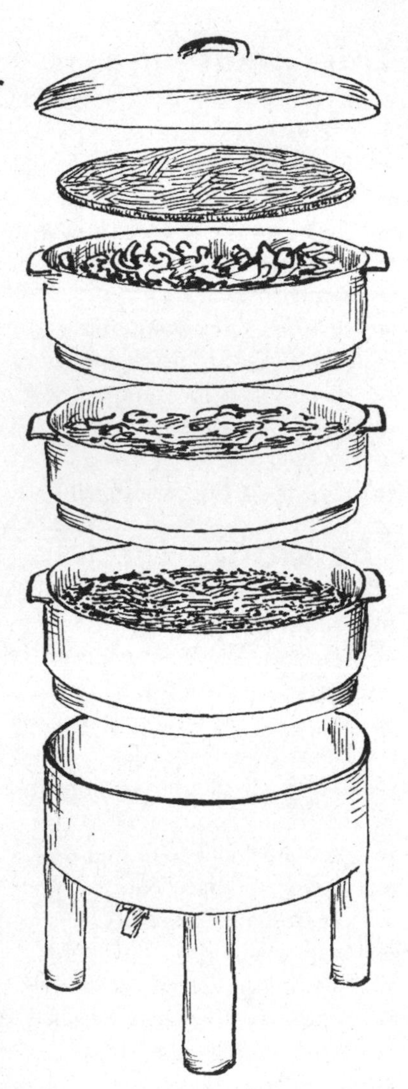

At a glance:

- ➥ Stacking Bin System (Upward Migrating Worm Composting Bin)
- ➥ Can handle kitchen waste of a 2-5 person family
- ➥ Made from recycled plastic
- ➥ Easier to manage than earlier plastic units
- ➥ Neat system
- ➥ May need additional harvesting by a dump 'n' sort method
- ➥ Raised off the floor on five push-fit legs
- ➥ 73 cm high (including legs) with a 50 cm diameter
- ➥ Tap for draining liquid
- ➥ Air vents incorporated into the bottom leg housing unit
- ➥ Widely available
- ➥ Comes complete with a full instruction manual

Stockists: **Wiggly Wigglers, Organic Gardening catalogue, West Country Worms** and **The Green Gardener.**

Worm Factory

At a glance:

- Stacking Bin System (Upward Migrating Worm Composting Bin)
- Two working trays with liquid collection sump and tap
- Additional trays can be added
- Space saving and rectangular in shape
- Raised from floor on four push-fit legs
- Close-fitting lid
- Can be supplied with coir brick, worms, lime mix, worm treat, and moisture mat
- Similar concept to the Can 'o' Worms, but simpler design
- Smaller than the Can 'o' Worms
- Measures 635 mm high, 550 mm wide and 400 mm deep
- Easy to assemble
- Easier to manage than previous plastic units
- Harvesting may need to be dump 'n' sort
- Comes complete with instruction manual

Stockists: **Wiggly Wigglers**

TerreNative

At a glance:

- Simple and transportable plastic tray
- Can stack several on top of each other
- Moisture wicking permeable clay liner
- Proofed against worms escaping - mesh-covered ventilation holes
- Tight-fitting lid
- 600 x 400 x 150 mm
- Place directly on level ground, no legs
- Can be supplied with coir brick, worms, lime mix, worm treat
- New design, imported from France
- Harvesting is the dump 'n' sort method
- Full instructions

Stockists: **Wiggly Wigglers**

Using worm compost 6

"Vermicompost outperforms any commercial fertiliser I know of..... I think the key factor is microbial activity..... Research that I and others have done shows that microbial activity in worm casts is 10 to 20 times higher than in the soil and organic matter that the worm ingests."

Professor Clive Edwards

Before you get stuck in ...

Personally, I have been digging my hands into soil, compost heaps and worm bins for years with and without gloves and I believe that my immune system needs this contact and exposure to various soil organisms, also that the use of antibacterial soap and the creation of a 'too sterile environment' would have an adverse impact on my health. However, it is a matter of common sense to wash your hands in hot water, using soap and a nailbrush, after working with worms and vermicompost.

Everyday practical uses of vermicompost

Once you have harvested your vermicompost, it is ready to use. The sooner you use it the better, as it will still be teeming with beneficial microbes. One of the real values of vermicompost is its sheer microbial activity, inoculating your soil or potting compost with beneficial microbes. Occasionally, there may be some inorganic material (such as plastic) or uneaten scraps left in the vermicompost, which should be removed, but generally speaking vermicompost is very clean and easy to manage. So let's look at how to benefit the most from your recycling efforts.

For house and garden plants

There seems little doubt that vermicompost is an excellent potting medium for greenhouse use, houseplants, general gardening activities and for larger scale plant growth operations in agriculture and horticulture. It can also be used as a planting soil for trees, vegetables, shrubs, and flowers or as a mulch, benefiting the soil when minerals leach into it. Such activities increase the rate of growth and health of plants and there is evidence that pest animals (such as nematodes) and diseases can be reduced. There is certainly room for more research into these aspects of vermicomposting in the future.

There are several ways in which vermicompost can be used. It can be used straight, mixed with soil or sand, put into planting holes, sprinkled on top of growing media, placed in seed furrows or made into a nutritious (for plants!) tea. A little goes a long way but you can never use too much, although it is too valuable to waste so you will want to use it sparingly. The pH is close to neutral. It won't 'burn', poison or rot plants; it is stable and non-compacting. Nutrient value will vary depending on what you have fed the worms but with the normal diversity of household waste, nutrients will be plentiful and well balanced.

Germinating seeds and seedlings

Vermicompost can be used to enrich commercial composts, for seed germinating or potting on. You can test for yourself which is the best medium - plant some seeds in a regular seed compost mix, some with the mix enhanced with vermicompost and some in pure vermicompost. You will end up with three groups of plants distinctly different in health, size and vitality. Vermicompost will prove its value before your eyes. Continue the experiment by potting the seedlings in a similar way using a regular seedling mix, then when the plants are of a good size, add vermicompost to the weaklings and watch them regain vigour. Just a small spoonful of vermicompost watered into a 10 cm pot will significantly improve the plant.

Pot plants

When potting-on established plants, mix vermicompost into the potting mixture, half and half if you can but even a small percentage will help. For your existing pot plants, sprinkle vermicompost over the top of the soil and water in. Do this every month and after a few additions, any worms hatched from cocoons in the vermicompost will start to thrive in the pot and continue the improvement for you.

If you don't fancy worms in a potted plant, for example if the pot is inside the house, simply liquefy the vermicompost and water the plant with the liquid. You won't get worms but you will get all the other benefits. Make this liquid by mixing one part vermicompost to about 10 parts of water. Shake vigorously and leave to stand for an hour or so in an open topped container. The sediment will settle to the bottom leaving a semi-clear, odourless liquid. This is best used while fresh. The liquid can be poured directly into your plant pots, or use a sieve, made out of an old pair tights, to filter out the sediment. You can also use the filtered liquid as a foliar feed for your plants. The sediment left after filtering may contain worm cocoons, which can be returned to your wormery or put on the compost heap. Or you could make a worm compost tea as covered later in this chapter and use that.

Flowers, fruit and vegetables

Add vermicompost whenever you can to flower and vegetable gardens. Simply sprinkle it on the soil and water in. Avoid digging it in, which could injure young worms and damage cocoons. The more you add the better the results will be but always water it in as this quickly releases the nutrients into the surrounding soil. Another method is to dig small trenches, fill them with vermicompost and then cover with soil or a mulch. Runner beans, French beans, potatoes, tomatoes, sweet corn and peas would all benefit from this. Add handfuls of vermicompost when you plant potatoes, and spread handfuls around when they are growing. Try to add vermicompost regularly.

Weeds from seeds

Some seeds may survive the vermicomposting process, as very little heat is generated. If you have been putting weeds in the worm bin, there is a chance some seeds may germinate in the garden. To overcome this, you may wish to consider liquefying the vermicompost or dig it in to trenches in the garden. You will see the weeds come up from the trenches but they will be easy to locate and easy to pull out.

Vermicompost tea. (VT)

It is produced by using stable, processed vermicompost and steeping it in water to release soluble nutrients and beneficial micro organisms. As with any brew, the method of brewing and the brewing material, is critical to its success. One of the most important factors here is oxygen. Commercial vermicompost tea makers use an Actively Aerated Compost Tea Maker (AACTM). These are increasingly becoming popular with domestic gardeners in the USA. Using an AACTM is far superior to any other method of making 'tea' yet invented.

Modern research by Dr. Elaine Ingham, a microbial ecologist of Soil Foodweb, Inc, has found that using vermicompost 'tea' on plants can:

- create a protective layer on leaves
- suppress diseases
- confer disease resistance
- improve plant growth
- reduce water loss
- deter foliar pests
- improve soil health and structure
- when used in pesticide damaged soils, it is proving very valuable in helping to re-inoculate them with beneficial microbial populations

For more information on vermicompost teas go to www.soilfoodweb.com/03_about_us/approach.html

To make vermicompost tea

You can easily make excellent vermicompost tea without the aid of an expensive machine. There are no hard and fast rules and you can vary the quantities to suit yourself.

Put 1 part of fresh vermicompost into a permeable container, a hessian or burlap sack, a pair of tights, an old carrier bag with small holes punched into it or even a sock. Put 5 parts of water into a suitable container then simply suspend the sack or old tights in the water. Agitate the bag several times by dipping it vigorously in and out of the water. Leave to stand for a short time and you are left with a rich tea.

If you are to spray the tea onto your plants you will need to strain the liquid by passing it through a sieve or another old pair of tights. You can use this liquid as it is or dilute it. Use the tea fresh as it loses its potency if stored. The left over, wet vermicompost can be used around plants in the garden or returned to your wormery or compost heap.

Lawns

Vermicompost tea is an excellent lawn conditioner. Spray or use a watering can to spread liberally over the lawn. To see the difference for yourself, treat just one half of the lawn. You will see the results in a reasonably short time.

Vermicompost can also be scattered onto the lawn and either watered or gently brushed in. If you have worm casts (little squiggly mounds) on your lawn, be grateful to your soil dwelling earthworms and brush them into your lawn.

Use vermicompost to enhance your soil

Many soils are cold and hard, clayey and devoid of life. Often this is because they lack organic matter, which as we know feeds the microbes. If you have a heavy, clay soil or a hard pan a few centimetres below the soil surface you can greatly improve it with vermicompost. First double dig the plot and create some trenches. Put a generous amount of vermicompost and manure or garden compost into the trenches. Water it in, cover with soil and put some more manure and vermicompost on top, watering this new layer with vermicompost tea. This will stimulate the soil microbes into a frenzy of activity. If the conditions are right, worms will hatch from any cocoons in the vermicompost in the trenches and on the soils surface, and make their way into the now enriched soil.

As soil is not their natural environment, most adult composting worms are unlikely to adapt beyond the trenches, but will continue to live there and produce cocoons. When the organic matter has fully decomposed and is no longer available for them to eat they will move away. By watering the soil with vermicompost tea, the micro-organisms will begin to condition the soil and start to make it 'worm friendly', which means gardener friendly too. The enrichment treatment may take several years to convert a cold, clay soil to a fertile and friable one, but it will happen.

Commercially available vermicompost and vermitea

There are several commercial companies that market vermicompost and document a range of improvements in the growth of many plants including vegetables, fruits, flowers and cereals. The concentrations of vermicompost that are recommended by such companies vary from 2 tons per acre for grapes, to 4 tons per acre for tobacco. For garden use the recommended rate of application is usually 2cm or so on the surface of the soil.

Some of the commercially available vermicomposts have been dried for ease of packaging and transportation etc. This negates one of its major advantages over most other commercially available composts - its enhanced microbial activity. The drying process will kill the microbes. The resulting nutrient stock and dried organic matter may have little benefit over other commercially available composts.

On the other hand, vermicompost tea is now also sold commercially, with very impressive results. This is a relatively new product and much research is being undertaken. It is made from stable finished casts that are, true to the tradition of English tea making, steeped in water to release the soluble nutrients and beneficial micro-organisms into the resulting brew. It is recommended for use as a foliar feed or to feed to plants via a watering can.

It could be said that commercially available products, like all 'active' vermicomposts, depend on strong aerobic microbial activity for maximum benefit. However for sale they are bottled or bagged (i.e. sealed in an airtight container, conditions that encourage anaerobic activity). The beneficial microbes are hardly likely to be flourishing when the product is opened. No doubt someone from a marketing department can explain.

Frequently asked questions 7

"The bones of dead animals, the harder parts of insects, the shells of land- molluscs, leaves, twigs, are before long all buried beneath the accumulated castings of worms, and are thus brought in a more or less decayed state within reach of the roots of plants."

Charles Darwin

Intruders

Q. There are too many flies in my worm bin. What should I do? How can I stop them?

A. There may well be several different species of fly in and around your worm bin, depending on bedding and foods you have put in. Most commonly you will find fruit flies, bathroom flies or mushroom flies. None of these are likely to be a problem in your wormery, except perhaps if the bin is indoors and there is a cloud of flies when you open the lid.

There are a number of methods you can use to try and reduce the number of flies in wormeries. To my mind we have to remember that a wormery is a complete recycling system and this includes all of the other decomposers, including flies.

Covering the food with a 2.5 cm layer of slightly damp shredded newspaper or better still vermicompost will soak up the moisture that the fly larvae need and deprive adults of potential egg laying areas, thereby reducing the fly population. Alternatively, opening the bin lid and allowing the surface to dry out will discourage the adults and kill the larvae.

To kill off the fruit flies some people actually freeze or microwave kitchen waste before adding it to the wormery. This seems a little excessive and a waste of time and energy.

Flies can be trapped in a jar with a holed lid, filled with a sweet fruity liquid and kept in the wormery. The flies will be lured in and be unable to get out again, but in my experience this is not very successful.

Biological controls can be used to eradicate fruit flies. Hypoaspis mites - a soil-dwelling mite, and a nematode worm Steinernema feltiae will attack fruit flies, but be wary as they may also consume organisms that are beneficial to the composting process. In laboratory experiments the earwig Euborellia annulipes was seen to eat wandering fruit fly larvae and pupae.

Q. I have ants in my wormery/compost heap. What should I do? How can I stop them?

A. If you have ants in your worm bin it indicates that it is too dry. Try gently watering the bin and then making sure it is kept damp. See Chapter 3 for further details. If your wormery has plastic legs, stand them in water or linseed oil if the legs are wooden.

Q. When I open the lid there are hundreds of red and brown mites all over the food, lid, walls and food waste and even on the worms. What are they? Will they harm the worms? What am I doing wrong?

A. These mites form part of the decomposition process. There are numerous species of mite, many of which might live in a wormery (see Chapter 3). As environmental conditions in a wormery favour one species over another, this can cause a community to bloom and multiply, apparently filling the bin. Eventually though, the tide turns to favour another community, and the previously abundant one declines. So, generally by doing nothing the bin will return to some sort of equilibrium.

Q. There are hundreds of tiny white creatures that jump. They are all over the food waste, lid and walls of the wormery. Are they a problem?

A. These are completely harmless collembola, commonly known as springtails and described in Chapter 3. They are very useful decomposers, and when conditions are favourable can appear to take over the wormery.

Q. Why are there hundreds of small white worms in my wormery?

A. These white earthworms are potworms (enchytraeids, see Chapter 3), a useful addition to your composting system. They look very similar to a newly hatched red worm, but do not have the red blood vessel running the length of their body. They can tolerate slightly more acidic and wetter conditions than the red composting worms.

There is a common misconception that a bloom of such potworms in the system indicates acidic conditions and that the pH requires management. It is a mistake to allow the management of the system to become pH oriented rather than composting and whole system oriented. The addition of a pH conditioner, often lime or calcified seaweed, can work against the larger composting worms in the system, which do best at pH5, quite acidic.

If you cannot bear to have potworms in your wormery the simplest method for decreasing their numbers is to allow the upper few inches of material in the system to become dry. These worms, like our red worms, have a high moisture requirement. When conditions are too dry, potworms will usually mass together in a ball and can be easily removed. Such a clump makes an ideal treat for robins.

It has been said potworms can be attracted by putting a piece of white bread, soaked in milk, on the bedding surface, though I've tried this and have not found it to be successful.

Q. Help! I have rats/mice/moles in my worm bin!

A. Larger animals are probably visiting your bin for food, but smaller ones may be looking for a nesting site too. Rodents like a reasonably dry place to nest, so keeping the unit wet will deter them.

Searching for food is harder to prevent. If it is an outdoor worm bed that they are visiting, lining the base with a 1 cm steel mesh and covering with a sturdy lid should keep them out. Mice and rats are opportunistic feeders. Mice particularly feed on seeds, grains, fruits, insects and their larvae, and worms will form a part of their diet, particularly near the breeding season when they need the protein. Buy a humane mice trap, one that I can recommend is obtainable from The Trap Man at www.traps.freeuk.com

Shrews will certainly make a dent in your worm population. Making the foodstuff unattractive to the rodents, for instance by mixing it thoroughly with a good carbon source such as wood chips, sawdust or cardboard, and keeping it all moist, should help. Once the foodstuff has started to be worked by the microherds and worms, it will be less attractive to the visitors.

A friendly cat may also be of help, as would a few rat traps if you are inclined to use them. I would not recommend the use of rat poison. Rodents have developed a tolerance to earlier poisons, so more potent chemicals have been developed. However, these linger longer in the rats and mice and have been found in lethal concentrations in the bodies of predators such as barn owls and weasels.

If you find voles in your wormery it should not affect the worm population, as voles are in the main vegetarian.

Worm issues

Q. I am going on holiday, what should I do about the worms?

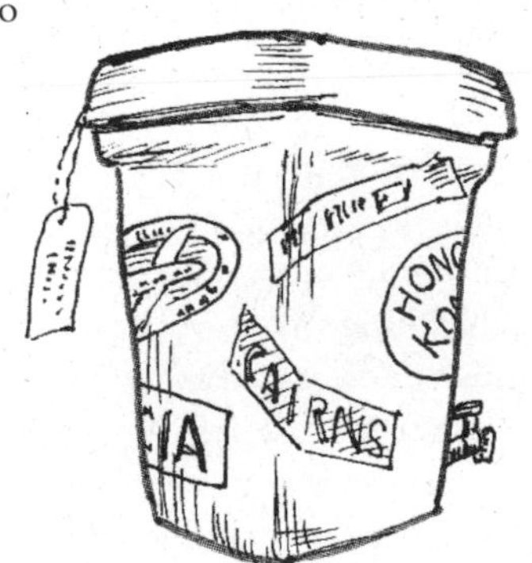

A. The good news is that you won't need to get a worm-sitter. Worms will happily re ingest their own faeces, as long as it contains organic matter teeming with busy microbes. I have been away on holiday for a full month and the worms survived. I did set them up for it before I left, adding soaked corrugated cardboard mixed in with their last feed to combat any potential anaerobic conditions. Adding soaked wood chips would also help.

Kelly Slocum, an enthusiastic and knowledgeable American vermiculture expert, undertook trials a few years ago with a couple of worm bins. Once they had reached peak efficiency she did not feed them, but watered with a hose. Although the individual worms were small, she saw little change in the total biomass of worms after 10 months. Therefore, as long as the moisture content of the wormery can be maintained, the worms will probably be OK.

Q. The worm bin is teeming with worms, will they spill over? Will I ever have too many worms in my bin?

A. You can never have too many worms in your wormery. Some worm species are prolific breeders, such as Eisenia fetida, commonly used for vermicomposting. Even so, worms will naturally regulate their own numbers, depending upon the food and space available and environmental conditions within the wormery. As the population increases, creating more competition for the available resources and the fouling up of those resources as they excrete, worms breed less often and do not grow as large. When the optimum conditions return, worms will resume breeding quickly again.

As mentioned earlier, it is the total biomass (weight) of worms that we are interested in for vermicomposting, and not individual numbers. If you feel you have too many worms you can put some in your compost heap where they may survive and speed up the decomposition process. But it would be better to use them to start a new wormery.

Q. What is the best temperature to keep worms working in my wormery?

A. The optimum temperature for composting worms is between 15 and 25°C inside the worm bin, but it depends upon which worm species is used. The most commonly used composting worm, E. fetida has an optimum temperature of 25°C. If we could keep this temperature constantly we may have the best results. But siting of the wormery, moisture content, and other factors would determine the temperature inside it.

Insulation from fluctuations in temperature is a good idea. Bringing wormeries indoors, somewhere like a garage, helps.

Q. My worm bin is outside, should I insulate it?

A. In a word, yes. Some commercial units already come with an insulation jacket. If yours does not, or you have made it yourself, an insulated bin would certainly help keep your worms alive and active during the colder months (see *Chapter 4 - Insulation of wormeries*).

Q. I have dead and dying worms on the top of the food waste. Some worms look like they have been sliced/pinched along their body, some are rotting, what is happening?

A. I have come across individual worms that look like they have been pinched or cut in several places along their body, leaving just the thin red central cord - very similar to a chain of sausages. Others have blisters, lesions or open sores along their bodies. It is a very sorry sight indeed.

There are several possible explanations, none of which have been completely proven. One is that it occurs when the worms have consumed something that can't be passed into the alimentary canal. This causes the blocked crop to swell, the worms move to the surface, become emaciated and die. Another is that their food contains too much protein or that there is salt contamination.

If you see 'pinched' worms like this in your bin, they are definitely dying, but nobody knows for sure what the exact cause. We can hazard a guess that the wormery has become too acidic and aerobic. I would overcome this problem by removing such 'contaminated' material, completely separating it and adding carbon rich dry material such as cardboard or wood chips which open up the heap and also dry it out. I would also only return half of such material into the wormery putting the rest into the outdoor compost heap. Leave the worms to settle down for a few days or even a week and then observe their behaviour.

If they appear to have settled down and are actively working the new material, then start to lightly feed again.

Q. Some of my worms have left the wormery - what am I doing wrong?

A. This phenomenon is commonly referred to as 'worm crawl'. Worm bins are designed to keep predators out, vermicompost in, but not to imprison worms. There must be some reason why they are leaving.

It is believed that worms are affected by rapid changes in barometric pressure, but this by itself is unlikely to be a cause of mass worm crawl.

If worm crawl occurs shortly after the setting up of a new wormery, or a new system, the worms may be simply reacting to the radical change in their environment. Using vermicompost as the bedding will reduce the likelihood of this occurring. Mature vermicompost is rich in a tremendous diversity of micro-organisms, which supply the vast majority of nutrients in the worm diet, and provides a comfortable environment for the worms so they are likely to stay. Other bedding materials offer little in the way of home comforts for the worms.

Other factors causing worm crawl could include anaerobic conditions, overheating, too dry or other unfavourable conditions.

The crawling behaviour generally stops after a few days but can last for a couple of weeks. Keeping a light on if indoors or leaving the lid open to sunlight will deter the worms, but alas not at night.

Q. When I open the lid, there are hundreds of worms clinging to it and on the walls. Why is this?

A. If you can see worms on the top surface of food waste in your wormery, that is a good indicator that they want feeding. However, when they are clinging to the lids or walls, it generally indicates that conditions in the bin are not to their liking and they are trying to move out. Being 'trapped' in the wormery, all they can do is cling to the lid or walls.

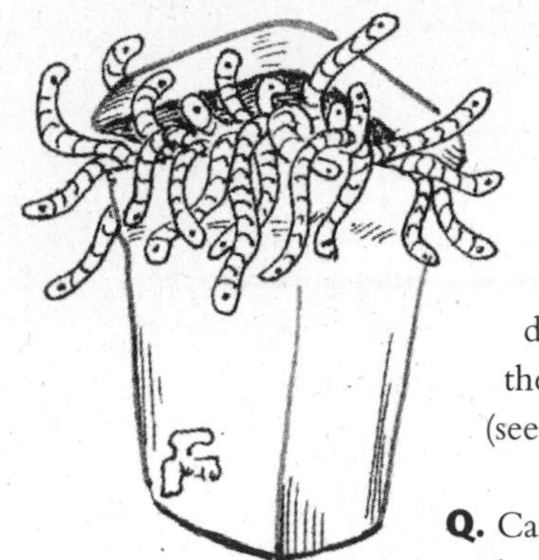

Very dry conditions can drive them to this behaviour, as there will be a little moisture on the lid and walls, especially in a plastic bin, but usually if it is dry they will move deeper into the bedding. Generally worms on the lid and walls indicate anaerobic conditions (see *Chapter 4 - What can go wrong?*)

Q. Can I use the worms from my worm bin to speed up the compost heap?

A. In a well-managed wormery you can provide the ideal conditions for your composting worms to not only survive, but also thrive. You can use these worms to speed up your compost heap, but the conditions are beyond your control and their advantage may be short lived - at worst they might be a tasty snack for something. Better off using them to start a new wormery.

Q. How long will the worms live in a wormery?

A. In the wild, on average, a worm may be lucky if it lives for a few months. But in captivity, safe from predators and extreme environmental conditions, there are records of Allolobophora longa worms living for more than 10 years, E. fetida for 4½ years and the common earthworm L. terrestris for 6 years. Hopefully your worms will live to a ripe old age.

Q. Will chemicals I use in the garden affect the worms?

A. This does depend on the chemical. There are many man made chemicals, particularly aimed at the domestic gardener, that may be harmful to worms. Indeed there was one chemical specifically formulated to kill worms in lawns. Thankfully that product has been withdrawn from the amateur market. But the garden centre shelves are still well stocked with chemicals that will have a serious affect on earthworm populations, reducing them drastically and for long periods of time, even after a single application. Methiocarb, a carbamate commonly found in pesticides and particularly molluscicides, is known to be highly toxic to earthworms and numerous other organisms.

Some chemicals may not directly kill the worms, but affect them in other unpleasant ways. They may, for example, affect their fertility, cocoon production, hatchling survival, growth, or induce weight loss. There is extensive scientific literature about the effects of pesticides upon earthworms, but very little knowledge as to the effects of other chemicals. Garden chemicals may also affect the micro-organisms that the worms rely on for their diet and your compost heap requires for the decomposition process.

To ensure that the microherds and earthworms in your wormery are not harmed or affected in any way, you have a few choices. Avoid unnecessary chemical applications, whether they be fertilisers or killers; read through the literature to select those chemicals most compatible with earthworm populations; or use none at all. As an organic gardener myself, I would recommend never using any synthetically produced chemicals. The same principle applies to your garden compost heap, where chemicals may also affect earthworm and microbe populations.

Q. I have heard that if you cut a worm in half, it will still live. Is this true?

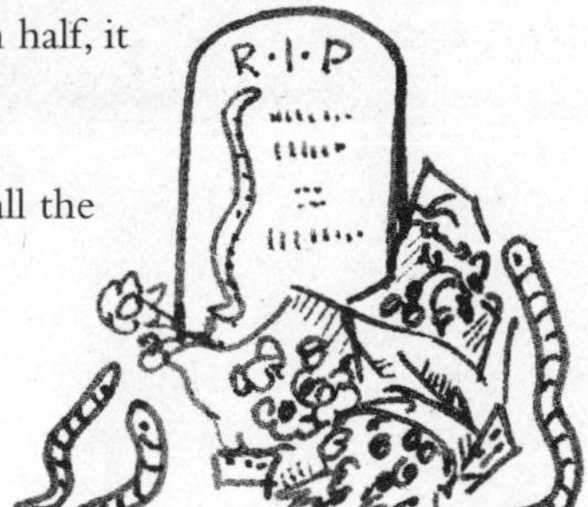

A. Just because worms look very similar all the way along, it doesn't follow that they are quite that simple. They do not like being cut in half and will die. It is true that sometimes when a worm is cut, both halves will wriggle around, but the result is not two surviving worms.

However, some worms, including those most often used in vermicomposting such as E. fetida, can survive and may regenerate a new head or a new tail if only their head or tail is cut off. This regeneration will depend upon soil temperature, age of the worm, worm species, number or segments severed and if the nervous system in the remaining segments remains intact. The regeneration occurs more readily with the tail end than the head. So, cutting them in half will not double your worm population.

Other management questions

Q. I have a new wormery. It seems to be taking a long time to produce vermicompost. Am I managing it badly? Why is it taking so long?

A. I have given many seminars and talks on composting, and I usually ask the attendees how long they expect to leave compost in their heap before it is ready to use. Most people say about a year, and a few even say two years. Thus I am always amazed when people say that they have had a wormery for 3 months and wonder why they cannot harvest it yet.

Most of the larger units I have had have needed at least a year before harvesting can begin. Why does it take so long? Imagine a packet of crisps, full to the brim. If you crushed the crisps to dust the volume would reduce dramatically and it would take many more packets of crisps reduced to dust to fill the original packet. This is what happens with your wormery. The microherds and worms reduce the volume and deposit it in a very fine form in thin layers, which gradually build up over a long period of time.

Depending on the system you use, harvesting too soon can deprive the worms of their bedding retreat. So have patience - it will be well worth the wait.

Q. My bin smells bad. There is a lot of liquid coming from the wormery. The worms are dying. The worms look like they are dead. They whole bin is in a mushy mess. Is it all right?

A. All of these situations arise as the result of the bin becoming anaerobic - see Chapter 4 on causes and the solution.

Q. There is a lot of brown liquid in the bottom of my wormery. It smells but I have been told I can use it as a liquid feed. What is it?

A. Some people call this liquid 'worm tea' or 'compost tea'. It is the by-product of anaerobic decomposition in your worm bin, and may contain all kinds of harmful substances. It should not be used as plant food - see

Chapter 4 on anaerobic conditions. Once anaerobic conditions in your bin have been addressed, the leachate can be poured back in (if the bin is not already too wet), or added to your compost heap.

A very useful liquid feed can be made from vermicompost, following the simple instructions in Chapter 6.

Q. What's the difference between aerobic and anaerobic microbes?

A. Their dependency upon oxygen. Those that require 15-20% oxygen are called strict aerobes. As the oxygen is depleted (around 8-9%) by, for example, rapid bacterial growth or if the oxygen is not available in the first place, facultative anaerobes (which can switch from aerobic to anaerobic metabolism) take over. Those that require the least, less than 2% oxygen are called strict anaerobes.

The microbes that work in such oxygen depleted environments produce by-products, gases and liquids, which we call leachate (see Chapter 4).

Q. My wormery looks OK, but it is quite smelly! Is it OK?

A. Smells don't always mean that the bin is anaerobic – some foods will naturally smell quite strongly as they decompose. Common culprits are potatoes and onions and brassicas such as cabbage, sprouts and broccoli. Bread will smell 'musty' and fruits will have a sharp sweet smell. Other foods will smell differently as they decompose, and all of these smells are subjective. But such smells should only last for a short while, until the worms and microbes have worked the foods, providing all the other factors are in place to assist them to do so. So don't automatically think "Oh no, my worm bin has gone anaerobic." And certainly don't add lime. Give it a chance to sort itself out before you rush into remedial action.

Q. I regularly turn my compost heap to aid oxygen. Should I do the same with my worm bin?

A. If all is well in your bin, the worms will do the turning themselves, opening up tunnels for oxygen to penetrate. If your worm bin starts to smell, it may indicate that anaerobic conditions are starting to dominate, in which case read the section in Chapter 4. Adding a good carbon source is the first line of redress: emptying the contents to turn, thereby disturbing and stressing the worms, is a last resort really.

Q. Should I add lime to counter acidic conditions in the wormery?

A. It is often said by companies selling worm bins and worms that pH management is critical, as the worms will die in acidic conditions, recognised by a foul smell. They will sell you a pH adjuster (i.e. lime, often in the form of calcified seaweed) to add to your bin and hey presto! your problem should be remedied.

However, acidity is generally not the cause of problem in your bin - it has become anaerobic and the only remedy is to add oxygen. How to do this is explained in Chapter 4. Acidity is one of the symptoms of an anaerobic wormery, and will be remedied when the problem is cured.

What this means is that pH control is not necessary in the worm bin, so don't waste your money. In the main, the management (or lack of it) will determine the pH in your worm bin. I certainly won't be wasting my time checking the pH.

Q. I have heard that lime can be beneficial to worms, why shouldn't I add it to the wormery?

A. Historically, farmers have limed soil to neutralise acidity and release plant nutrients such as magnesium and calcium.

Research indicates that soil-dwelling earthworms may benefit from lime added to soils, and adding lime to acidic soils usually favours worm populations. It is known that the addition of lime to certain soils increases the amount of plant growth, which, in turn, results in an increase in the amount of organic matter, which then becomes available for the worms to utilise as food. It is also probable that certain earthworms have a need for calcium as part of their diet.

However, most of this research has been undertaken adding various forms of lime to mineral soils, usually in agricultural fields. Whether these benefits apply to adding lime to rotting vegetation/kitchen waste is highly questionable, where different microbes are at work and in the main different species of worms.

On the other hand, adding lime to wormeries will suddenly change the pH dramatically, completely altering the environment for the inhabitants of the wormery. The microbes may be forced into dormancy, or even killed. This will be counterproductive to the vermicomposting process, to say the least.

Q. Is all right to use tap water in my wormery?

A. Although the worms probably prefer non-chlorinated water, I have used both tap water and rainwater for many years with no obvious ill effects.

Chlorine is added to tap water in order to prevent bacteria growth, so there might be an argument for using rainwater in preference.

Food questions

Q. What quantities of food should I give the worms?

A. This goes back to the decisions you made before you bought or made your bin. The size of your bin determines how much food you should put in - the rule of thumb is of $^1/_2$ kg per $^1/_{10}{}^{th}$ square metre of surface area (one pound per square foot) each week.

Having said that, the rate at which the food is processed will be affected by so many factors: what is being fed; what condition it is in; density of worms; population of worms; environmental conditions; temperature and moisture content within the bin; the bin itself; and so on.

There is no correct answer, but there are certainly wrong answers. Many people tend to overfeed their worm bins, which creates problems, particularly anaerobic conditions. They may have been used to disposing of their garden waste by piling it high on the garden compost heap. This method does not work with worm bins. Little and often is the secret.

The most important thing to remember is, if there are worms on the top of the bedding looking for food you can feed them, if you cannot see hungry worms, do not feed them.

Q. I have been told not to compost food waste as it will go acidic.

A. It is true that food waste will become acidic when it begins to rot, but in an aerobic worm bin (or compost heap) it is not in any way a problem. At the beginning of the natural decomposition process, especially with vegetable and cooked food waste, slightly acidic conditions are created by the microbes themselves. This is not detrimental to the environment in the bin, it is part of the cycle and will soon sort itself out. However, if starved of oxygen (as mentioned earlier) the bin will remain acidic and the worms will suffer.

Q. Am I managing the wormery correctly? There's mould all over the food waste. Will it harm the worms?

A. Mould will not harm the worms at all. This is part of the natural cycle of decay and recycling. There are microbes at work here, slowly decomposing the foods. You are observing a visible process in the form of fungi and mould, which thrive in dark, damp conditions. Like the visible creatures such as mites and worms, when conditions are right the fungal community thrives. Once the microbes have sufficiently decomposed the material, eventually the worms will begin feeding in this microbially rich area.

If anything, mould on the food may be an indicator that perhaps you are overfeeding your system a little, or that there is low worm activity for the amount of available organic material, as can be the case at cooler times of year.

Q. Can I add food waste to the worm bin in winter? It seems to take a long time to rot down.

A. I have been doing exactly that for many years. Certainly, activity will slow down, but the more we can do to assist the system the better. I keep my units in a brick-built garage, and they are made of wood, which is quite a good insulator. However, I have found that in plastic units the worm activity decreases so much that it is best to stop adding food, putting it on the compost heap instead until you can see some worm activity.

Outdoors it would be more of a challenge to keep the worms busy all year round, but there are people who vermicompost in Alaska during the winter. As well as a good layer of insulation, you need to keep the carbon and nitrogen balance right.

Q. Can I feed anything to the worms to increase their numbers?

A. Researchers have found that if you feed bullock or horse manure to earthworms, they will produce more cocoons than if you feed them sewage sludge, farmyard manure or straw. They also increase in weight proportionally more when fed animal manures rather than other foods. In my opinion this is a far better 'worm treat' than the commercially available ones, which are pretty expensive compared to manures

Q. When I have a lot of food waste, should I reduce it in the blender?

A. No, it will only contribute to making your wormery anaerobic. It is popular in the USA, where they call it 'blenderising'. As with shredding garden compost material, blenderising will increase the surface area for the microbes to work upon, thereby increasing their activity and numbers. However, as in the garden compost heap, particle size will affect the oxygen flow. In a bin with reduced levels of oxygen, aerobic microbes will suffer and anaerobic conditions will prevail.

It has been found that such liquefied feeds may actually favour microscopic nematode worms to the detriment of our friends the composting earthworms.

Personally, I would not 'blenderise' any foods. You might want to reduce 'heavy duty' things such as corn on the cob, which do take a long time to decompose in a worm bin. But it is not necessary on a domestic scale, and just means another thing to wash, and wasted electricity.

Q. Should I put less food in the worm bin while it is becoming established?

A. There are a number of factors to consider here. The worms themselves are not the only creatures doing the composting. The microbes also play a huge part. As the material passes through the gut of the worms the numbers of microbes is greatly increased. The bedding material is actually a

huge reservoir and breeding bed for microbes as well as for the worms.

So, if you start your wormery with fresh vermicompost as bedding, not only the worms but also the microbes are raring to get going breaking down the foods. You can then add fresh foods straight away, starting with a thin covering and gradually building up. If you are starting a new batch of vermicompost with the compost including worms harvested from your previous batch, the same would apply.

However, if you are supplied with, for example, coir or hemp blocks or you make the bedding from other materials, then the worms have to consume this to convert it to suitable bedding material. Adding food waste would only prolong this bed making process. So, in this case, it would be wise to wait for the worms and microbes to start consuming and decomposing the bedding material before adding anything else. How long this would take would depend on how many worms you added at the beginning. and how much non-vermicomposted bedding material is supplied.

Q. Can I compost food waste at a school or in my own home?

A. In general you are free to compost any kitchen waste. Where catering or household waste contains meat, it must be composted in a premises approved by DEFRA under the Animal By-Products Regulations 2003. The only exception is for catering waste composted on the premises on which it originates, provided that the material produced is used only on that premises, and that livestock are not kept on the premises. This means that householders may compost their own kitchen waste on their own domestic compost heap. It also means that premises such as schools, hospitals or prisons may compost their own kitchen waste for use on that premises, provided livestock are not present.

Q. I have bought some plastic bags that claim to be biodegradable - can I dispose of them in my wormery?

A. If they are biodegradable bags, and not just degradable, you might be able to put them in the wormery. It depends on what they are made of and how much plasticiser is in the material, if any. Generally the first ones that were manufactured did not vermicompost very well. I suspect newer ones may break down better in a wormery, albeit at a slower pace than if put in an aerobic compost heap.

I would recommend you try it, but monitor the first bag to see how it breaks down - perhaps tie a piece of string to it so you can retrieve it periodically. I suspect it would really do better in a compost heap.

Vermicompost questions

Q. What will the nutrient value of my vermicompost be?

A. Vermicompost is not a simple, inert thing, and we cannot say that it contains this, this and this because it was once grass, bananas, paper, tea bags or whatever. Obviously, the nutrient levels of castings vary considerably with the type of material being fed to the system.

However, there is more value in vermicompost than a simple list of nutrients - because of its microbial activity it benefits the soil and the plants growing in it in a multitude of ways. Its main benefit is not in providing a balanced set of nutrients for the growing plants to utilise, but improving the whole environment for growth - soil life, soil texture, water-holding capacity, nutrients made available to plants etc. While many plants show immediate, dramatic response to the use of castings, in an equal number of situations the benefits are long-term health improvements, which are not necessarily measurable in relation to the addition of vermicompost.

The levels and forms of nutrients are constantly changing in fresh vermicompost, while the worms are continually breaking down organic matter and consuming microbes and their faeces. Eventually, when the microbial activity finally dies down, the levels of nutrients would be constant, but the main benefit would have been lost.

Q. What will the pH of my vermicompost be?

A. When worms process the food they do not affect the pH, so ultimately the pH of the vermicompost will depend on the pH of the bedding and foods you supplied. As the food is broken down by microbes it will become acidic for a time, but gradually this will return to a fairly neutral pH. Many organic wastes tend to be on the alkaline side of neutral (i.e. above pH7), and your vermicompost is likely to be, too.

Q. Will the worm bin kill weed seeds?

A. This is unlikely. In a fully thermophilic compost heap (one that generates considerable heat) most weed seeds will perish. A worm bin is totally different and such temperatures would drive the worms away. Worms may consume weed seeds. They will pass through the worm and out the other end, completely unharmed, viable and encased in their own mound of 'ready to use fertiliser', similar to blackberry seeds after passing through a blackbird. In the wild, weed seeds are ingested by worms and deposited elsewhere.

Q. I have seen vermicompost and vermicasts offered for sale - what's the difference?

A. Technically speaking, vermicompost is a mixture of casts and organic material. Casts, if they are 100% casts are exactly that, pure casts with little or no organic matter left as it has all been processed through the worms. To be sure exactly which was in the bag, the company would have to do a certain amount of research.

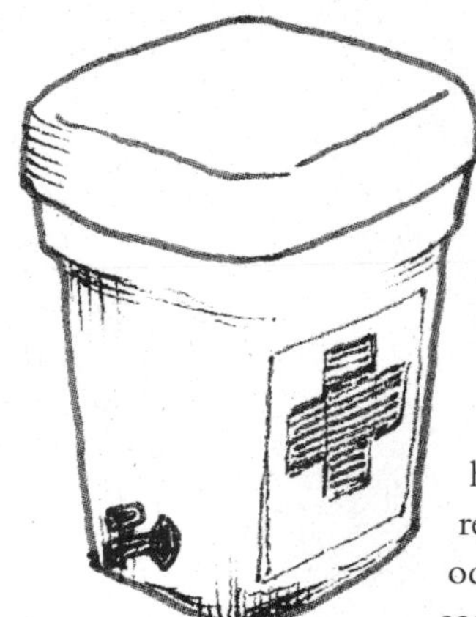

Q. Are there any health issues with regards to vermicomposting and wormeries?

A. Any potential health concerns relating to composting and vermicomposting may be dependent on the individual concerned, the material being vermicomposted and also the unit's location.

People suffering from asthma and those who have compromised immune systems or an allergic response, should be aware that there are naturally occurring moulds and fungi in worm bins and compost heaps, and they may need to take special precautions. The use of a damp mulch mat on top of the contents of the worm bin or compost heap may well minimise spores being released into the air.

Common sense dictates that potentially risky foods such as meat/fish/poultry, bones, carcasses and cat/dog and exotic bird faeces should not be put into worm bins. It is known that composting worms can destroy human pathogens, but we cannot guarantee that if they are any pathogens they are removed. More research is being undertaken with regard to pathogen control using worms and the early results are encouraging.

I have been digging my hands into soil, compost heaps and worm bins for years with and without gloves without any problems. Personally, I subscribe to the argument that my immune system needs this contact and that exposure to various soil organisms keeps it functioning at its best. I believe there is more a risk of something being passed on to me from a fellow human than I do from a compost heap or a worm bin!

It is a common sense, basic hygiene precaution that after digging garden soil or turning the garden compost heap we should wash our hands in hot water, using soap and a nailbrush.

The varied roles and wider uses of worms 8

"Good to eat, and wholesome to digest, as a worm to a toad, a toad to a snake, a snake to a pig, a pig to a man, and a man to a worm."

Ambrose Bierce

Those of us who farm worms appreciate how useful they are, but their value extends far beyond domestic recycling. This chapter will look at the many uses of worms, as well as a few people who have done invaluable work focusing on their benefits to mankind.

Clive A. Edwards. Father of vermicomposting?

Clive Edwards is recognised as a world authority on earthworms. An Englishman who studied at Bristol University, later joining the Rothamstead Experimental Station in 1960, he concentrated his studies on the soil and the effects of agrochemicals. He recognised at an early stage in his career the huge potential of using earthworms to decompose organic matter and waste material and has done much important research into vermicomposting animal manures and food wastes. His book Biology and Ecology of Earthworms has run to three editions, and was the first serious work to be published on earthworms since Darwin's book on the subject in 1881.

His research led him to conceive the continuous flow system (see Chapter 4). He later developed completely automated reactors, each capable of processing thousands of tonnes of waste per year. The continuous flow system is now commonly used throughout the world for medium and large-scale vermicomposting processing. It has more recently become available to the domestic user. Without doubt, his research has opened the way for the vermicomposting industry and the domestic vermicomposter to easily recycle a waste product into a valuable resource.

Global vermicomposting today

Vermicomposting is now practiced all over the globe, but it is much more widespread, both commercially and domestically, in Australia, New Zealand and the United States than elsewhere. Vermicomposters around the world include schools, universities, colleges, supermarkets, farmers, householders, rabbit breeders, apartment dwellers, office blocks, airforce bases, gardeners, horticulturists, camp sites, educational establishments, museums, fruit markets and even hospitals. The record here in Great Britain is dreadful. We introduced E. fetida into most if not all of those countries outside of Europe. The Australians even used worms and composting techniques at the Sydney Olympics 2000, where organisers aimed to recycle 80 percent of the expected 5,500 tonnes of Olympic-generated waste. Can you imagine that at Wembley or Cardiff? What a missed opportunity to not include the Wembley Wonder-Worms in the rebuilding plans!

Community composting

However, there has been a growth in awareness of environmental issues and landfill problems. Many individuals and small organisations have looked for a better way of disposing of their rubbish. All over Britain, community composting organisations are springing up. One such is the Community Composting Network (CCN), which offers advice and support on numerous matters concerning composting to communities wishing to compost waste. It has a library amd display materials etc for loan, and provides training events and newsletters for its members.

Large-scale vermicomposting in Britain

Although worm farms have existed in the UK for a number of years, they have in the main been restricted to growing worms for the fishing bait market.

Perhaps the best-known academic advocating large-scale vermicomposting of wastes here in the United Kingdom is James Frederickson of the Open University. He is recognised as a leading authority on using earthworms as part of a waste management programme. He was instrumental in developing a huge experimental site for vermicomposting waste at the Worm Research Centre in East Yorkshire.

Prior to this centre, no in-depth scientific research had been undertaken on the potential of worms used outdoors here in the UK, bearing in mind our climatic conditions. Most of the research had been in laboratories where optimum conditions can be artificially controlled. The Worm Research Centre was established on the site of an old worm farm, to address the issues of maintaining optimum environmental conditions for native worms, outdoors.

Worms as 'night soil men'.

In days gone by, before the advent of the flushing toilet, 'night soil men' were employed to take away human manure. With more research, worms may become the new night men. Over the years there have been several attempts to vermicompost sewage, or biosolids as it is now called - mostly unsuccessfully. Scientific research and published knowledge on the vermicomposting of biosolids is limited, although small-scale laboratory experiments have been undertaken.

Currently the two main methods of dealing with biosolids (after treatment) are to pile it into heaps and treat as compost, or spread the waste over the soil surface. With either method worms are then left to work the material.

Great improvements could be made in onsite processing and disposal of large quantities of sludge using worms. Research has found that passing biosolids though the guts of worms reduces the odours associated with such material and stabilises it. The passage through a worm also reduces the levels of certain pathogens present in the biosolids. New research, undertaken by Dr. Elaine Ingham of Soil Foodweb Incorporated USA, shows that worms are able to 'clean' foods supplied to them. Dr Ingham studied the presence of two micro-organisms, Escherichia coli (a human pathogen) and Fusarium oxysporum (a plant pathogen), in the food supplied to worms. Both could be detected in the gut of the worm near to the mouth, but further inside the worm, and in the worm casts, neither was present.

This is in contrast to the effect that worms have on soil microbes, which increase in numbers as they pass through a worm. Researchers believe that the conditions inside a worm's gut are ideal for soil micro-organisms to thrive, even more so than conditions in the soil, and this may be partially due to the large amounts of mucus and water that earthworms secrete into their guts. Particular microbes have been seen to proliferate in different areas of the worm gut, for reasons we do not yet understand.

The conditions inside a worm may be ideal for soil micro-organisms, but plant and human pathogens, and other microbes that do not naturally live in soil, are adapted to very different conditions. The reasons they die inside a worm may be various - they may starve, suffer unfavourable environmental conditions, soil microbes may predate on them - we can only guess. For more information about Dr. Ingham's work and soil life go the soilfoodweb.com website

It would seem worms could have a massive environmental impact, turning a problematic waste product into one that could be used to fertilise crops, condition the soil, enhance crop production, even be bagged and sold to the public.

For more information about human pathogens and vermicomposting see the excellent work 'Achieving Pathogen Stabilisation using Vermicomposting' by Bruce Eastman in BioCycle Nov 1999 or visit the website below to read this article and many more useful ones www.jetcompost.com/refrence/

Worms and the spread of microbes

Although earthworms have been found to reduce the numbers of human pathogens in biosolids, they have been accused of spreading microbes, not always beneficial ones, through soils from one area to another. There are many kinds of pathogens, causing diseases in humans, animals or plants.

As well as pathogens, worms are known to consume and distribute the nitrogen-fixing bacteria that form mutually beneficial relationships with clover and other leguminous plants, mycorrhizal fungi, and most plant communities. Worms are known not only to distribute such beneficial bacteria but also to enhance their numbers.

Worms as tools for environmental management

Land Reclamation

The many positive aspects of having worms in the soil have led to the practice of improving poor soils by inoculating them with earthworms. Reclaimed soils, often from urban or other highly managed land, have been found to be more productive when the natural colonisation by worms is supplemented by adding adults, juveniles and/or cocoons from colonies cultured domestically. There have been several instances where worms have very successfully improved poor soils, such as open-cast mines and landfill sites, and there is considerable promise for a role for earthworms in restoration schemes in the future. However, there is still much to learn about the process.

The types of species added are crucial to the success of such projects. Worms used in vermicultural practices are often not suitable for reclaimed soils - they thrive in compost or manure heaps and soils of very high organic matter, and this is not usually the state of soils being reclaimed.

Another potential application that is currently being investigated is using vermicompost as a growth media (i.e. a soil) to be directly placed on reclaimed land. Of course the very likely, subsequent colonisation by 'wild' worm populations could enhance and improve such habitats.

Farmland reclamation

We tend to think of landfill sites and mines when we talk about land reclamation, but farmland can need improving too. Farmland is subject to many stresses worldwide, such as soil erosion, compaction by heavy machinery and over-treatment with potent and toxic agrochemicals.

Worms incorporate organic matter into soils, recycling plant nutrients and making them readily available to the growing crops. The walls of vacated worm burrows are rich in plant nutrients and are ideal channels for plant roots to enter, which in turn allows the plant to utilise the plant nutrients and aerate the soil. The organic matter in the top layers of the soil acts like a sponge, soaking up valuable water as it seeps into the soil using the vacated worm tunnels and the spaces between the plant roots as channels. If water cannot find a way into the soil it remains on or near the soil surface and is lost. Or in the worst cases it can cause serious erosion by washing the topsoil away leaving the subsoil exposed.

So worms are beneficial to farmland, but they are under threat. Tilling and ploughing the soil is known to drastically reduce worm populations, as is the use of molluscicides (slug killers), particularly methiocarb. Farmland that has had recent applications of methiocarb or recently been ploughed will have very few healthy worms living in the soil.

Land that lacks worms, lacks water-holding capacity and natural nutrients in a readily available form for strong healthy plants and is therefore more likely to suffer from soil erosion. Adding organic matter to the soil and inoculating farmland with worms will enhance the natural health of the soil. Recent evidence has suggested that farmland with large earthworm populations produces more stable long-term yields.

Modern research to date has found that using worm compost 'tea' can suppress and provide disease resistance to plants. When used in pesticide-damaged soils, it is proving very valuable in helping to re-inoculate them with beneficial microbial populations.

Vermicompost by prescription?

Bearing in mind that vermicompost is the result of the activity of worms and associated microbial communities acting on whatever organic matter they are provided with, we know that it is highly variable in composition. To a large extent 'what goes in, comes out' but this is of course dependent on exactly what is living in the bin and the prevailing conditions.

However, the composition (or 'castings value') can to a large extent be manipulated to the needs of the user, by providing worms with different foods. For example, it has been suggested that worm casts intended to amend soil for planting a fruit orchard need to be dominated by microscopic fungi. Therefore, casts need to be produced by worms fed with a high proportion of carbon material (e.g. leaves, wood, paper, cardboard). Vermicompost intended for adding to soil that will be growing annual flowers needs to be dominated by bacteria. Such high bacterial casts are produced with foods higher in nitrogen (e.g. green wastes, manures). Some day we may be able to feed worms a particular food so that they are able to produce vermicompost that deals with a specific situation for us.

Worms as toxic land indicators

It is now widely accepted that agrochemicals can accumulate in the soil. It is also accepted that pesticides can also affect species other than the ones targeted. One way that this can happen is by the consumption of the target species by their predators.

Another side effect of agrochemicals can be altered behaviour. For example predatory beetles were found to prefer to consume slugs that had been poisoned by slug pellets instead of consuming live healthy slugs, thus enabling the healthy slugs to continue damaging crops.

Another example of a pesticide affecting 'non target' species was highlighted by Rudd investigating the now notorious chemical, DDT. In this case the chemical was being used to control the beetle responsible for Dutch elm disease. It was found to have accumulated in robins, with lethal results. Painstaking research found that the robins had eaten earthworms that were full of DDT. How did it get into the worms? Literally, it had dripped off the trees onto the soil and leaf litter which the worms had consumed. Rudd found that worms can accumulate up to ten times the concentration of DDT in their bodies than is present in the surrounding soil, before suffering ill effects - in this case they are termed 'biological concentrators'.

As well as some pesticides, worms can accumulate various heavy metals from polluted soils in their body tissues. This is either by ingesting the soil with the pollutant, or through their skin just by being directly in contact with the contaminated soil. One result of this close connection between worms and the soils in which they live is that they can be used as an indicator of the 'health' of soils. Using worms or other animals as 'bio-indicators' can help the monitoring of soil condition by measuring the toxic chemicals in their tissues.

Toxicity testing

To be fair, it is recognised that worms are very valuable to soil and pesticides must be tested to ensure minimal impact on non-target animals including several species of worms, as well as other beneficial organisms such as beetles and pollinating bees. Toxicity tests involve directly immersing the worms in solutions of chemicals or feeding them contaminated food.

Land bioremediation

As well as showing that land may be contaminated, earthworms can be used to cleanse contaminated soils. This was found by researchers, such as Trevor Piearce, from Lancaster University working on spoil heaps from abandoned copper/arsenic and tungsten mines. Such contaminants would prove fatal to most earthworms as well as other animals, but the researchers found that Lumbricus rubellus had colonised the spoil heaps.

Somehow these worms manage to survive in the highly toxic slagheaps. This makes interesting work for researchers. Did the worms develop a resistance to these toxins, was this resistance a physiological or behavioural adaptation or could it be a genetic one? When fresh worms (again L. rubellus) were introduced to the spoil heaps from a different site they fared very badly and were unable to build up the same resistance.

Many interesting areas for research have opened up, and currently scientists are investigating such areas as:

- the behaviour of earthworms in relation to contaminated soils;
- how earthworms are able to live with and detoxify contaminants;
- earthworm/microbe interactions in contaminated soils;
- the potential for using earthworms/microbes in the bioremediation of contaminated sites.

The future of modern soil science

In a handful of garden soil, there may well be more biological diversity than there are animals and plants in the tropical rain forests, yet those few centimetres beneath our feet are the least studied place on Earth. Now at last it is attracting the attention of more scientists. £16.8 million has been set aside for a five-year research study of the soil, undertaken by researchers from seven countries and backed by the United Nations. Life in the soil, an ecosystem in its own right, microbes, fungi, mites and other organisms, including worms, will all be studied. These soil dwellers have yielded some of the world's most important antibiotics, including streptomycin and penicillin. The search is on for strains of nitrogen-fixing bacteria that could save the millions of pounds currently spent on artificial fertilisers.

To give just one example, in one area of tea plantations in India, soil had degraded and yields had stalled even after heavy applications of fertiliser and plant growth hormones. Scientists re-introduced our friends to regenerate the soils. This simple act produced an increased yield of up to 280% on some plantations and subsequently increased profits.

The hall of fame in worm research has only a handful of members: Clive Edwards, Jim Frederickson, Trevor Piearce and Kevin Butt from the UK and in North America worm taxonomists Sam James and John Reynolds (an ex policeman, like myself) and researchers such as Elaine Ingham and Bruce Eastman. We need more dedicated worm researchers like these. And then there are the great worm advocates like Kelly Slocum and Mary Appelhof (who sadly died recently) both from the USA. So let's have you budding worm researchers out there. It's an interesting area to research, with huge potential.

Vermicomposting has potentially so much to offer in dealing with our ever-increasing mountains of organic waste and our ever-decreasing topsoils throughout the world. The greatest benefits could be seen in remote areas or poor regions of the world.

The future of modern science lies with our children and their education. We have mentioned elsewhere how worm composting can be an amazing topic to teach children important skills, sciences and concepts, together with valuable life lessons that , once taught, can be used at home. Besides, kids love worms.

Worms can be pests – well, nothing is ever perfect!

Firstly, let me stress that the benefits of worms to soil structure and fertility are many, and that earthworms almost always benefit the soils they inhabit. However, they can have a negative impact on their environment.

Agriculture

Despite their benefit to agricultural soils, under certain circumstances earthworms can be pests of crops. The behaviour of some worms to drag material on the surface down into burrows has led to damage occurring to crops such as lettuce. Their burrowing can cause damage to the delicate roots of seedlings, and they have been found to damage the roots of rice in paddy fields. Also, surprisingly, if produced in sufficient quantities, worm casts can interfere with farm machinery. Generally, however, earthworms are not considered to be a significant pest of crops in any way.

Grasslands

There are places, such as golf greens and other sports playing fields, where casts on the surface are not welcome, and some people object to casts on their lawns. Also worms attract that bane of the groundsman - the mole. For these reasons they are controlled in some areas.

Worms as prey (and a source of protein)

Vertebrate predators

It is no surprise that animals as abundant, defenceless and nutritious as worms are preyed upon by bigger animals. Worms are an important component of the food chain for many of their predators, which include mammals, reptiles and birds. Worms can be purchased from pet stores and from catalogues for lizards etc. and even for feeding to your wild bird visitors.

Invertebrate predators

Creatures smaller than worms also eat worms - there are flies that parasitise them, mites that parasitise worm cocoons and micro-organisms that kill worms.

Worms are also eaten by their fellow invertebrates (animals without backbones). Rove and ground beetles, centipedes, leeches, some species of slugs and ants, and even other earthworms all benefit from the protein in a worm. And the high profile invertebrate predator that has wreaked havoc on earthworm populations in many countries including both the UK and the US is the introduced New Zealand flatworm.

New Zealand Flat Worm, NZF (Arthurdendyus triangulatus)

This worm was accidentally introduced to the UK through the plant trade with New Zealand. It was first recorded in Northern Ireland in 1963 and has since spread to Scotland and Northern England, but there have also been sightings in the South East.

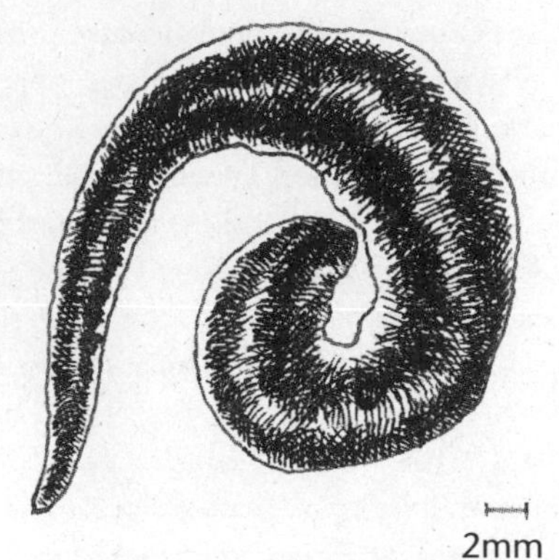

As the name suggests, the worm is flat. Adults have a purple/brown upper surface with buff-coloured margins and underside. It is usually about 1 cm wide and 5 cm long, stretching to 10-15 cm when it is moving. It has a smooth but very sticky skin. If you think you have found one, see the website www.defra.gov.uk/planth/pestnote/flat.htm for further details.

The NZF is mainly active at night, spending the daytime hiding under logs, stones, paving stones, inside plant pots, and other objects. In fields, they tend to stay near the margins, where there are places to hide. They move in a similar way to our earthworms, secreting mucus to ease their passage through the top few centimetres of soil and over the surface. Earthworms are caught either on the soil surface or in their burrows. The flatworm wraps itself around its prey and releases enzymes that reduce the worm to a soup, which it then eats.

These predators may have a considerable impact on the environment far beyond the scope of reducing earthworm populations. We know that worms contribute to a healthy soil and soil structure, and some scientists suggest that the NZF will have sufficient impact on worm populations to alter it. There is potential for reduced drainage and the subsequent water logging, an increase of surface water run-off, pollution and flood hazards, and a decrease in agricultural productivity in certain areas. Also wildlife that rely on earthworms as part of their diet will be effected. It has been suggested that some species of worm may be more susceptible to predation than others and that there would be an alteration in the species of worms in the soil after the NZF invaded a particular area.

A greater knowledge of their biology is necessary to control the spread and impact of these devastating creatures. Certain beetles have been reported to attack them, but on the whole they are not prey to wildlife in this country. There must be stricter regulatory control for importing and exporting plants and similar materials to prevent such a disaster from happening again.

Needless to say, if you find an NZF in your garden do not touch it as the mucus may cause skin irritation. Refer to the DEFRA action notes above. Also, keep an eye out for more - where there is one you will probably find others.

Source of protein

As a source of protein, worms are comparable to meat and fish. When alive, worms are 75-80% water, but protein makes up 65-70% of their dry weight. Pigs, fish, chickens, rats and shrimps have all been fed on processed worms, which might have been air-dried, freeze-dried, oven-dried, acetone-dried, liquidised, blanched or mixed with molasses.

Some research has been conducted to investigate the potential of worms as protein sources for cattle. Our composting worm, E. fetida, was used to determine the viability of farming worms for protein. Compared to tenderloin beef, which costs £9 ($16) per kg to produce, loin of worm costs £12 ($21) per kg, so it is not really a viable option. An important consideration here in the UK is our recent history with BSE, which may have been caused by the feeding of infected sheep carcasses to cattle. Cattle should eat grass and although I'm sure the occasional worm finds its way into a cow by accident, feeding them worms feels rather like making the same serious mistake.

I'll have mine rare!

In the past, earthworms have formed part of the human diet. Worm farmers in Korea hope to reawaken the appetite, claiming that eating worms helps normal blood circulation and cures thrombosis. After a ten-year struggle with legislators, worms were added to the official livestock list in 2004, allowing the farmers disaster relief from the state and loans for setting up.

Although it requires a big investment to build the proper housing to cover worm fields, worm farming is a lucrative business. A one-acre farm can produce more than 5 metric tons of worms annually and is perhaps the most lucrative agricultural endeavour on a per square metre basis.

But the government refuses to register earthworms as a food or medicinal product, on the basis that the idea of eating them is disgusting. At the same time Koreans enjoy eating fried silk worm maggots, from stands on street corners.

Recent legislation

After the outbreak of the Foot and Mouth epidemic, the UK Government introduced the Animal By-Products Regulations (ABPRs) 2003. These aim to prevent animal by-products from presenting a risk to animal or public health through the transmission of disease. They state that where catering or household waste contains meat, it must be composted in a premises approved by DEFRA. Green waste composting is not regulated by the ABPR.

There is an important exception that allows small-scale operators such as household gardeners, schools and community groups to compost their waste. Catering waste can be composted on the premises on which it originates, provided that the material produced is used only on that premises, and that livestock are not kept on the premises. This means that householders may compost their own kitchen waste on their own domestic compost heap. It also means that premises such as schools, hospitals or prisons may compost their own kitchen waste for use on that premises, provided livestock are not present.

The regulations are specifically aimed at people operating authorised plants using animal by-products (storage, incineration, processing, composting etc.). They are likely to affect all those that deal with animal by-products. So if you have really got the vermicomposting bug and want to set up a large-scale scheme, you would be well advised to read these regulations. You would also need to speak to the Environment Agency who offer technical guidance on composting operations, regulate the activity, issue licences and enforce the numerous other Acts and Statutory Instruments involved in large scale Licensed Waste Management systems)

• Copies of ABPRs are available from The Stationery Office (previously HMSO) or on their website: www.legislation.hmso.gov.uk/si/si2003/20031482.htm

The Composting Association

If you are really taken with vermicomposting, you should know about the Composting Association. Formed in 1995, it works 'to promote the environmental and economic benefits of effective composting and compost use.' Although mainly attracting businesses involved in the composting industry, such as commercial compost producers, local authorities, trade suppliers, and academics, it does have individual and student members. The Association develops and runs workshops for those in the industry and promotes 'Composting Awareness Week'.

Worms in schools

Our society has reached a stage when it can arguably be classed as one of consumers first, social citizens second and ecological citizens when it suits us.

Disposal of food waste has long been a thorny problem. Most people just bin it. Recent research suggests that one third of all food is thrown away. If we add to this the increasing use of disposable cutlery, cups and plates, fast food outlets, microwaveable meals on disposable trays, and so on, we have to ask ourselves what are we teaching our children?

Hopefully, at some time during their stay in school, children are taught about waste and its impact on the environment. Vermicomposting can be an invaluable tool in that educational process. Understanding and practising worm composting, seeing where it fits into the waste cycle is a valuable lesson in turning the philosophy of 'reduce, reuse, recycle' into a working reality.

Children are generally fascinated by worms and earthworms are easier to care for than most animals. A wormery with a perspex window, such as the 'Waste Buster', would make an ideal unit to have in schools, as students are able to observe the biology, ecology and morphology of earthworms in their natural surroundings. The students would have a clear view of adult and young earthworms, cocoons and the community of other creatures that make wormeries their home. In an ideal world, all schools should have worm bins to convert their waste to useful compost - sustainability in action.

Some enterprising pupils may wish to develop their business skills by selling the worms to their fishermen fathers or the vermicompost to gardening parents.

Some ideas for wormy activities in the classroom

There are numerous activities available for teachers and pupils to explore together; these are just a few suggestions...

- **What's in your rubbish bin?** List items in bin. How much can be reused, recycled, composted? What is left? What happens to it? Where does it go?
- **What's in your lunchbox?** Leftovers from students' lunches become dinner for the worms, teaching students that 'waste' can be a resource and even the lowest of creatures (the worm) can make a difference.
- **Worm menu.** What can worms eat? What can't they eat?
- **Environmental awareness**. What is waste? How does our community dispose of it? How much food do we throw away? What is recycling? What can be recycled? What does Sustainable Development mean?
- **Maths**. Waste audits, charts, graphs, reproduction rates, measurements of food consumed, height/weight of plants, length of worms, etc.
- **Science**. Do plants grow well in vermicompost? What else lives with the worms? What makes up vermicompost? Food web of the compost heap. Who was Charles Darwin?
- **Biology**. Do worms have eyes? A skeleton? How do they move? What is the swollen band called? Do they have bristles? How do worms live, reproduce and react?
- **IT skills.** Searching the Internet for worm-related stories, facts, data handling, contacting schools/organisations re vermicomposting

A very interesting teachers resource pack for primary classes can be downloaded as a Portable Document Format file (PDF) by visiting http://www.interplayuk.co.uk/home_page.html
clicking on the Living World logo, then highlighting Worm World.

Appendix 1

A Worm's Menu

We know that what worms eat is actually microbes, but they do like their microbes to be eating something themselves. All of the materials that are listed here, unless stated, can be added to your worm bin (or to a garden compost heap). There are many things that could go in the wormery, many of which we don't normally think of as 'food', that are just thrown away - a waste of an excellent resource. This list is not exhaustive (or it would fill more than just one book). As a general rule, all uncooked and cooked vegetables, fruits, their peels, and grains can be fed to worms. Although worms are able to consume meat, fish, poultry and poultry bones, we tend not to add these to our worm units.

With all worm food, remember that 'little and often' is the rule. If the food is fine or very sloppy avoid suffocating the wormery by adding too much. Alternatively, you could mix such food with a bulking agent, like woodchips, cardboard or leaf mould.

Aquariums - Why waste a useful resource? When cleaning out your fish tank, why not add the waste water to your wormery or compost heap. Not only does it contain certain algae and manure but also it will moisten the bin.

Beans - fresh, dried, cooked or baked.

Bread, pastry, biscuits, cereals and flour - Worms can eat bread, but some people have found problems that, when providing bread alone in large quantities, it can start to ferment and even 'pickle' the worms. If bread dries out, it can also be a problem as it just lies on the surface looking unsightly and smelly as the butter or margarine decomposes. Make sure it stays moist, and add woodchips, newspaper or cardboard at the same time.

Be careful when adding flour as too much will just form a nasty, slimy mess. Little and often is the answer.

Breakfast Cereals - all, including their boxes

Cakes - like us, worms just love cakes!

Cardboard (Egg boxes, corrugated boxes, cereal boxes, cardboard trays, loo rolls). Worms can be very happy in cardboard alone, as long as it has been shredded or reduced into fibres. However, it is much better to mix it in with green, organic material to get a good balance.

Some cardboard and paperboard (the material used, for example, in cereal boxes) has a wax coating to protect the board from moisture. This coating may take a little longer for the microbes to process, but eventually even wax is digested by microbes.

Casseroles

Cheeses - all; cottage, mouldy, old, new, even Danish blue.

Chocolate and chocolate products - but why waste it on worms?

Citrus rinds - Citrus peels contain an essential oil called d-limonene. This oil is widely used as an industrial and domestic degreaser. It can irritate the skin of worms, and can be fatal in quantity.

Small amounts of citrus peel will not affect your worms. Like all organic matter, it will be broken down slowly by the microbes. Worms will stay clear of it until the oil has been broken down and the peel itself has been substantially softened. Adding large amounts of citrus peel to your wormery may cause problems. Ordinarily, I just throw my citrus peels into the wormery and let the worms get on with it.

Coffee grounds, husks, leftovers and filter papers. These will eagerly be consumed by worms. Mixed in with other material, they will cause no problems, but do not suffocate the bin with a blanket covering of coffee grounds. Caffeine is OK for worms but it seems to deter slugs, so perhaps you should save your grounds for the slugs.

Comfrey - roughly chopped, an excellent natural source of potash.

Compost - Fresh compost is rich both in organic matter and microbes, and your worms will love it. As the microbes use up the organic matter, their activity and numbers decline, and after a period of time it will be much less appealing to your worms. Some materials take longer to decompose than others so would remain attractive for longer. So, old garden compost may be of little value to the worms as food, but it could be useful as a bedding substitute when first starting a wormery.

Old compost is useful added to your wormery as a top layer over a layer of food, especially if the food is fruit. It should limit the number of fruit flies. Another advantage of adding compost to your wormery is that it may introduce other decomposers such as woodlice, slugs and snails. You may or may not be pleased about this.

Cotton, wool, rags, old clothes, sheets, etc. In time the worms will work their way through all your old clothes. Don't forget they cannot eat your old nylon jumpers, lycra running shorts or anything that is not made from natural fibres.

Corn on the cob kernels - Worms love to hide down the middle of these and eat them from the inside. As they are hard they can take some considerable time to decompose - chopping them up or crushing them will speed the process up a bit.

Dried foods - I have used dried food on a few occasions. Once it remained completely dry on top of everything and never decomposed. The other time I overdid it, covering the bedding with a 5 cm layer. I watered it in which effectively filled up all the air spaces and suffocated the worms. So, if you want to add dried foods, just add a little and mix it with everything else.

Eggshells - I have always added eggshells to my worm bins. Sometimes I crush them, sometimes I oven dry then crush them and sometimes I forget. Either way when I harvest the end product, invariably there are still fragments visible in the vermicompost. Shells that have been boiled are said to be more persistent.

Some people add eggshells to their worm bins to provide calcium for the worms. Small amounts of calcium may be needed for healthy worms, and it is probably useful in the worms' gizzards to assist in grinding food. I add it to my bin, but not for some complicated reason - just because we sometimes eat eggs.

Feathers from pillows, duvets etc - they will take some time to be fully broken down but the worms will get there in the end.

Flowers - dead, cut or dried.

Fruit and fruit peelings, cooked or uncooked.

Garden fertilisers such as sterilised bone meal, dried blood, hoof and horn. I personally do not use them, as I would be wasting my money adding them to the contents of the wormery when they could go straight onto the soil. I do prefer to use dried seaweed or a liquid seaweed in my wormery and compost heaps. A definite no-no is super phosphate or sulphate of ammonia as they are lethal to worms. One researcher actually recommended sulphate of ammonia be used on golf courses to rid them of worms.

Garden waste, weeds and green manures – very small quantities only, so as not to generate heat. Best use them on your garden compost heap.

Grass cuttings – You can add small amounts to your worm bin, mixed with newspaper or shredded cardboard. However, the most beneficial way is to leave them on the lawn or use them as a mulch around garden plants where they will break down and add organic matter and nutrients to your soil.

Fresh grass cuttings are exceedingly rich in nitrogen and have about 75% moisture content. This balance of material, its structure and size, makes it the ideal material to encourage a massive growth of aerobic microbes. In a compost heap, this activity will generate a lot of heat very quickly. Such temperatures will be too hot for the worms. This energy is very soon used up and the heap will quickly turn sour and anaerobic - not good for the worms either. Once this initial heating stage is over, the 'pre-composted' grass cuttings could be transferred to the worm bin (but not a great huge layer) and they would probably go down quite well.

Grass cuttings that have dried out or chopped hay and straw are a different matter. The fast acting nitrogen will have been spent, so this is valuable as a material to soak up excess moisture and as a food source for the microbes in the wormery. Again, too much would be a bad thing, so moderation, as in all things.

Hair (pet and human) - Hair can be difficult to deal with in large quantities - it will form a mat and shed water, making it very difficult to break down. Hair is worth adding to the bin, though, as it has a high nitrogen content - add small quantities mixed with cardboard, newspaper and other carbon items.

Kitchen and toilet roll tubes - just cardboard, a wormy favourite.

Leaves from trees - In small amounts leaves can be safely added to your wormery. Earthworms do eat them, and different species can differentiate between the various tree leaves and may even show preferences for some. Beech and oak leaves and pine needles contain tannins, which can put the worms off. Soaking freshly fallen leaves for a day or two and discarding the water may speed the process up a little. Leaves that have lain on the ground for some time will have had some of the tannins leached out, and may be more palatable to the worms.

Leaves that have dropped from trees are not very rich in nutrients, as the tree has reabsorbed them before casting off the leaf - the leaves visibly change colour when these nutrients are withdrawn. Adding leaves to your compost or wormery will capture the remaining nutrients, but they will take a long time to decay because they contain very little nitrogen.

If you have a lot of leaves I would make leaf mould, which the worms love, and add that to your wormery. I often throw beech and oak leaf mould, which is itself at least a year old, onto the top of the worms' food to stop fruit flies. This will add fibre and aid moisture retention, helping to make the end product very crumbly and friable. The worms seem to like it, as do the woodlice.

Leftovers from the table - worms love them but be careful not to add too much at a time.

Magazines and glossy paper - These can be vermicomposted, but I have found that they do take longer to decompose than non-glossy paper and I prefer to recycle them through my local paper recycling facility. Although vegetable based inks may now be used, there is some discussion about heavy metals being present in some inks.

Manure (farm animal dung) - Without doubt this is the magic ingredient to be used in wormeries, but not when too fresh. Although good vermicompost can be made without it, the extra effort in obtaining and preparing it is well worth the bother. I have found that my worms really thrive when a layer of aged horse manure is added to my worm bins, every six weeks or so. It acts as a tonic for them and I would thoroughly recommend this treatment. Manure contains microbes in their billions, busy consuming the manure. They will readily start decomposing your waste. Several researchers have found that the addition of animal organic matter (particularly pig manure) greatly increased the cocoon production of earthworms over those that just consumed fresh plant material.

However, fresh raw animal manure and any bedding accompanying it, especially if it contains a lot of urine, is a definite no-no for a wormery. As well as urine, it might contain heavy metals or concentrated salts, which would harm the worms, and there may be pathogens present in the manure.

Manures should be allowed to age ('pre-compost') for 10-15 days in a separate heap. The manure will probably undergo a hot, aerobic decomposition, which will break down weed seeds and kill pathogens, evaporate off excess moisture and expend the urine (nitrogen) in the breakdown of the carbon material in the heap. Pre-composting allows time for nasties such as vermicides, heavy metals, artificial fertilisers and salts to be leached away. All of this conditions the material for faster worm processing and reduces the potential for overheating in the wormery.

The concentrated and bagged manure that you can purchase from garden centres and the like has been pasteurised and dried prior to being packaged or bagged. This process ensures the destruction of any potential pathogens/diseases etc. that may have been present in the manure, as well as all the beneficial microbes. Adding it to your wormery will never have the beneficial effects of pre-composted manure.

Mayonnaise - Worms will consume mayonnaise, but in large quantities it can cause smells, so bury it and mix it with carbons such as wood chips.

Meat, fish, poultry and bones etc - Following the Foot and Mouth and BSE scares we have had in this country I would not recommend adding meat, fish and chicken, poultry etc. to your worm bin. Obviously small quantities of items containing meat will find their way into the system, and that is unlikely to cause problems. I occasionally put chicken bones in my wormery, keeping them well buried. They decompose completely fairly quickly.

In the USA there are many people who vermicompost these materials and have done successfully for years on a domestic scale.

Nail clippings - Your toe and fingernails can be successfully vermicomposted. Now that is real recycling.

Nettles - Worms like nettle leaves, which is no bad thing because they are full of valuable nutrients. I often chop them up and throw them on top of my worm bins to give the worms a bit of a treat.

Nuts, nutshells and sunflower husks - all these products will be eaten but nutshells will take a little longer.

Onions and onion skins - Although they may take a little longer than other foodstuffs to disappear in the worm bin, they do go, and so does their smell. However, if I had a lot I would compost them before I put them in the worm bin. Just as a precaution I would recommend no more than 10% of the total of fresh food be onions or their skins.

Paper and newspaper - Not only can newspaper be consumed by worms, it is an important addition to a wormery. Also, cardboard, copy paper, fax paper, laser printer paper, office paper, and even those telephone directories, which of course should be ripped first (no mean feat, I must admit). I would go so far as to say that if you are only adding food waste to your wormery, it is essential to add paper too. The food waste will be mostly nitrogen and it needs to be balanced with carbon. You may find your unit becomes smelly and anaerobic if you do not get this balance right. Also, paper soaks up any excess moisture and aids the free passage of oxygen in the worm bin, especially if shredded or ripped.

In the past there were concerns about adding printed paper to compost heaps and wormeries, but most printers have now switched from coloured inks containing heavy metals (i.e. toxic) to vegetable based ones, and these are harmless to worms. Black inks are usually made from hydrocarbon or petroleum based compounds, which are broken down in the composting process.

Pasta - cooked and uncooked.

Peat and coir - These are both used widely as bedding and as food for worms. Personally, as an environmentalist, I cannot advocate their use. We should be using home-grown, recycled bedding that is reasonable and sustainable, rather than transporting coir thousands of miles or destroying our last few irreplaceable peatlands.

Pet faeces - Hamsters, guinea pigs, and other such vegetarian pet manure can safely be added untreated to your wormery. Rabbit manure, has a very high nitrogen content and if there are large quantities, it should first be soaked in water before adding, along with the bedding to leach out any ammonia, urine and salts. Discard the water or use it directly onto your compost heap. Pet bedding can be used in a wormery provided it is not the synthetic type. However, the faeces of animals such as dogs, cats, pot-bellied pigs, parrots, budgies and the like are a different matter. These animals can carry pathogens and parasites which can cross into the human system. We cannot yet guarantee that they are destroyed in the vermicomposting process.

Vermicomposting of pet wastes is something that has been practised for many years in Australia and in fact is promoted by some Local Governments there as an alternative 'disposal' method for these wastes. One company there actually manufactures and markets a small wormery as a 'doggy-do disposer'. I am aware that people do compost and vermicompost pet faeces and use the 'compost' as a fertiliser on non-vegetable areas, but for your health's sake, do not use cat or dog faeces in your worm bin!

Pet food - Dry pet foods, that are cereal based and do not contain meat, will be OK if added in small quantities with other foods. They may be high in salt and protein, both of which could cause problems for the worms if large quantities were added.

Post-consumer waste - the stuff we scrape off our plates or throw away after it has been prepared or cooked.

Potatoes - baked, boiled, chipped, dauphinoise, even uncooked potato peel.

Pre-consumer waste - basically means materials used before we eat or use them, e.g. peelings, discarded lettuce leaves, cabbage stalks and the like.

Puddings - see chocolate

Rice, rice cakes and rice products

Salad and salad waste - in fact, any greenery

Salt - in quantity will kill your worms, but in food that has been cooked with salt it is sufficiently dilute and will do them no harm.

Seaweed, fresh, dried or liquid - Fresh seaweed contains too much salt and should not be added to your worm bin. The salt could be washed away, but it would be a huge effort.

Research has shown that beneficial soil bacteria are enhanced when liquid seaweed has been applied as a drench, and it increases plant root growth. I water liquid seaweed onto my bins from time to time as I believe it will also stimulate microbial activity in my worm bins, although I have no personal scientific data to back this up. I do occasionally sprinkle dried seaweed onto the surface of my worm bins - I have found that the worms enjoy this, actively rising to the surface for it.

Seeds from weeds, plants, the bird feeder or the kitchen - Ask any vermicomposter about tomato seedlings! Although worms will ingest small seeds they certainly do not digest all of them, and some will be viable when they are expelled in the casts. So if you put seeds in your worm bin there is no guarantee that they will be killed and you may find the plants coming up all over - in the top of the worm bin, in your house plants and seed trays. Simply pull them up and throw them back in with the other food wastes, they will soon be broken down.

Soil - Although soil per se does not form part of a composting worms diet, a handful or so in the wormery aids decomposition as it introduces beneficial microbes. The small particles of rock present in the soil are useful to the worm, which has a gizzard and needs grit to grind its food.

Spoiled and mouldy food - don't waste it, give it to the worms.

Tea bags and tea leaves - Worms will eagerly consume tea leaves. I have found tea bags disappear more quickly if they are ripped prior to adding to the heap, as the worms can then get to the tea leaves without having to wait for the paper to be broken down by the microbes. In winter, I have also found it beneficial to squeeze excess liquid from the tea bags before ripping and putting in the bin - they do hold a lot of fluid and large numbers of soggy tea bags can keep your bin very wet. Alternatively, add newspaper to soak up any excess liquid. Tea bags and tea leaves, of course, can also be used as a mulch around shrubs.

Vegetables and their peelings, cooked or uncooked

Vermicompost – This is an excellent innoculant for starting a new wormery or a new batch, and makes the best bedding for worms. So do not use it all up on your plants.

Wood, sawdust, shavings and chips - Decaying wood naturally forms part of the diet of worms, so small pieces of wood will be accepted in the heap. Fine sawdust will tend to exclude air, though, so mix with other foodstuffs. Wood chips are very useful in compost heaps and wormeries, retaining moisture and opening up the material to allow the passage of oxygen. They provide a balance to the carbon in kitchen waste. Wood shavings are a halfway house between sawdust and wood chippings.

Timber from coniferous trees contains hydrocarbons such as terpene, an important ingredient in turpentine. I would not pour turps onto my worms, and would not add fresh sawdust or wood shavings from such trees. I recommend pre-composting them first, or adding them with manure that has itself been pre-composted. Cedar wood contains oils that are highly resistant to microbial action, so if you add it to your bin it may take a long time to break down.

Worm fattener/treat - There are companies that market worm treats or worm fatteners. The ingredients may include bran, powdered milk, wheat flour, ground pellets and a form of lime, possibly calcified seaweed (a non-sustainable product). Personally, I feel there is really no value to worm fattener whatsoever. We are not interested in individual worms and their size, it is the total weight of them that matters. Why give the worms a 'treat'? While they are eating the treat they are not working the other material in the unit, i.e. your waste food.

Yeast and brewers waste - I have heard that a few large-scale commercial vermicomposters use yeast as a worm treat. It is sprinkled on the top of the feed stock and it attracts them in droves. It could be used to attract worms to a new worm bed, or out of the compost being harvested. Used sparingly it should present no problems.

Stockists of worms and worm bins:

British Eco
Northgate House, St Mary's Place, Newcastle upon Tyne NE1 7PN
Tel: 0191 209 4161
www.britisheco.com

Green Gardener
1 Whitmore Wood, Rendlesham, Suffolk. IP12 2US
Tel: 01394 420087 Fax: 01394 420064
www.greengardener.co.uk

Nurturing Nature Ltd
1 Banbury Drive, Great Sankey, Warrington WA5 1HW.
01925 452819
www.nurturingnature.co.uk

The Organic Gardening Catalogue
Riverdene Business Park, Molesey Road, Hersham, Surrey KT12 4RG
Tel: 0845 130 1304 Fax: 01932 252707
http://www.organiccatalog.com/catalog/

Original Organics
Unit 9 Langlands Business Park, Uffculme, Cullompton, Devon EX15 3DA
Tel: 01884 841515
www.originalorganics.co.uk

The Recycle Works Ltd
Unit 1 Bee Mill, Ribchester, Nr Longridge PR3 3XJ
Tel: 01254 820088
www.recycleworks.co.uk

West Country Worms
Collaton Farm, Blackawton, Totnes, Devon TQ9 7DW
Tel / Fax: 01803 712738
www.westcountryworms.co.uk/

Worm City
128 Kingfisher Way, Ringwood, BH24 3LN
Tel: 0871 2116569
www.wormcity.co.uk

Woods Farm Organics
Woods farm, The Common, Earlswood, Solihull B94 5SQ
Tel: 01564 702963
www.woodsfarmorganics.co.uk

Worm Hotel Ltd
Blythe Cottage, 44 Littler Lane, Winsford, Cheshire, CW7 2NF
Tel: 01606 592145
www.thewormhotel.com

Worms Direct UK
Drylands, Ulting, Nr Maldon, Essex, CM9 6QS
Tel: 01245 381933
www.wormsdirect.co.u

Contacts

Community Composting Network (CCN)
67 Alexandra Road, Sheffield, S2 3EE
Tel: 0114 258 0483 or 0114 255 3720
Email: info@communitycompost.org
www.communitycompost.org/

The Composting Association
Avon House, Tithe Barn Road, Wellingborough, Northamptonshire NN8 1DH
Tel: 0870 160 3270 Fax: 0870 160 3280
Email: membership@compost.org.uk
www.compost.org.uk/dsp_home.cfm

HDRA, The Organic Organisation
Ryton Organic Gardens, Coventry, Warwickshire CV8 3LG

Tel: 024 7630 3517 Fax: 024 7663 9229
www.hdra.org.uk

Soil Foodweb Incorporated USA
http://www.soilfoodweb.com/

The Worm Research Centre
http://www.wormresearchcentre.co.uk/

A UK based worm forum
http://www.recycleworks.co.uk/cgi-bin/ubbcgi/ultimatebb.cgi

Web sites

http://www.jetcompost.com/reference/index.html
http://www.planetark.com/dailynewsstory
http://www.soilfoodweb.com/03_about_us/approach.html
http://www.wormdigest.org/forum/index.cgi
http://www.eco-logicbooks.com

References and selected bibliography

Appelhof, M, (1997), 'Worms eat my garbage', Flower Press, Michigan, USA - distributed in the U.K. by eco-logic books

Appelhof, M, (1997),'Worms eat our garbage', Flower Press, Michigan, USA - distributed in the U.K. by eco-logic books.

Eastman, B, R, (1999) 'Achieving pathogen stabilization using vermicomposting' BioCycle, The JG Press Inc. Emmaus, PA USA. Nov pages 62-64.

Edwards, C S, & Burrows, I. (1988) 'The potential of earthworm composts as plant growth media'. In Earthworms and Environmental Waste Management. Eds. C.A. Edwards and E.F. Neuheuser. SPB Academic Publications, The Netherlands, pp 211-220.

Edwards, C A & Bohlen, P J, (1996),'Biology and Ecology of earthworms', Chapman & Hall, London, UK.

Langan, A.M., G. Pilkington and C.P. Wheater (2001). Feeding preferences of a predatory beetle (Pterostichus madidus) for slugs exposed to lethal and sub-lethal dosages of metaldehyde. Entomologia Experimentalis et Applicata, 98, 245-248.

Lee, K E, (1985)'Earthworms: Their ecology and relationships with soils and land use'. Academic Press. Sydney, Australia.

Payne, B, (1999),' The Worm Cafe : Midscale Vermicomposting of Lunchtime Waste – A Manual for Schools, Small Businesses and Community Groups' Flower Press, Kalamazoo, Michigan, USA.

Rudd, R.L (1964)'Pesticides and the living landscape', Faber, London, UK.

Savigear, E, (1992),'Garden pests and predators', Blandford, London, UK.

Sims, R.W. & Gerrard B.M. Earthworms: Keys and notes for the identification and study of the species. The Linnean Society of London and Estuarine and Brackish Water Sciences Association. London, UK.

Wallwork, J A, (1970),'Ecology of soil animals', McGraw-Hill Ltd, Berkshire, UK.

Wheater, C P, (1999),'Urban Habitats', Routledge, London, UK.

Those books still in print can be ordered from eco-logic books

We will leave the last words of this book to the late Mary Appelhof who in May of 2002 posted the following news on the Worm Digest forum:

"Squirm of Worms joins Gaggle of Geese, Pod of Whales, Pride of Lions as a collective noun"

So now we know what to call our wormeries - squirm hotels